Doin

DOING IT

Women Working in Information Technology

by

Krista Scott-Dixon

SUMACH
PRESS

WOMEN'S ISSUES PUBLISHING PROGRAM

SERIES EDITOR: BETH MCAULEY

LIBRARY AND ARCHIVES CANADA CATALOGUING IN PUBLICATION

Scott-Dixon, Krista, 1973-
Doing IT: women working in information technology/
Krista Scott-Dixon.

Includes bibliographical references.
ISBN 1-894549-37-6

1. Women computer industry employees. 2. Information technology. 3. Women — Employment. I. Title.

HD6073.C6522S36 2004 331.4'81004 C2004-904280-7

Edited by Beth McAuley
Cover and design by Elizabeth Martin

Sumach Press acknowledges the support of the Canada Council for the Arts and the Ontario Arts Council for our publishing program. We acknowledge the financial support of the Government of Canada through the Book Publishing Indusry Development Program (BPIDP) for our publishing activities.

Printed and bound in Canada

Published by

SUMACH PRESS
1415 Bathurst Street Suite 202
Toronto Ontario Canada
M5R 3H8
sumachpress@on.aibn.com

Contents

Acknowledgements
7

INTRODUCTION
Women's Work in Information Technology
11

CHAPTER 1
Women's IT Work in Context
31

CHAPTER 2
The Struggle for Skills
63

CHAPTER 3
Great Promises versus Material Realities
105

CHAPTER 4
New Work versus Same Old, Same Old
128

CHAPTER 5
Looking Ahead
164

Appendix A
Data on Paid and Unpaid Work
191

Appendix B
Interview Participants
197

Notes *203*

Glossary *229*

Selected Bibliography *233*

Index *241*

ACKNOWLEDGEMENTS

Writing a book is a collective effort even if only one name appears on the cover. It would have been impossible to conceive of this project, never mind complete it, without the assistance, guidance and feedback from many people. Alaina Hardie deserves top billing as technical advisor, since in her position as Supreme Geek Goddess, she has provided nearly unlimited information about the IT industry. This includes plenty of juicy gossip about various corporations (she anticipated, for example, the spectacular implosions of many dot-coms), the proper syntax for vi commands, anecdotes about funny IT co-worker antics and all sorts of real-world critiques about my half-baked theories on women and technological work. She patiently read several drafts with unfailing cheerfulness and didn't mind when I phoned her at work to whine that my FTP wasn't working.

This book grew from much of the work I did for my doctoral dissertation, and I am most grateful for the wise direction of my PhD committee: Norene Pupo, Pat Armstrong and Ester Reiter. Norene, in particular, offered timely counsel, assistance and nurturance throughout the project at critical intervals in its evolution. The gentle and thoughtful interrogation provided by Pat, along with the incisive critiques suggested by Ester, assisted the development and growth of the research as well as the researcher. Carl Cuneo, Livy Visano and Margo Gewurtz all devoted careful attention to reading the work, and asked excellent, provocative questions. To a scholar, there are few more valuable gifts than readers who engage enthusiastically with one's writing, and I am thankful for all of their ideas.

As a young academic, having supportive mentors is crucial to one's success. York University seems to be a magnet for many of Canada's foremost feminist researchers in the field of gender and work, and by happy accident that is where I have been, first as a student, then as a cog in the academic machine, since 1991. I am grateful for the ongoing informal mentorship and guidance provided by Linda Briskin, who turned me on to Women's Studies when I took her class twelve years ago. Leah Vosko, a feminist political economy powerhouse, has provided an intellectual foundation and enthusiastic support for my work, enabling me to add layer upon layer of theoretical complexity to my understanding of the field. Her research has been fascinating and innovative in its conceptualization and representation of precarious employment. In terms of career and intellectual development, I am utterly in Leah's debt. The very sweet and always supportive Barbara Crow somehow found time in her manic schedule to read through a crude draft of the work and provide detailed remarks on it. Heather Dryburgh from Statistics Canada helped me make sense of a bewildering array of statistical evidence, and reassured me that I was on the right track.

Many people at the university advised me to look for a university-based academic publisher. But as soon as I met the folks at Sumach, I knew that this small women's press would be a much better choice for me, and for the kind of stories I wanted to tell. Beth McAuley managed to distill a nugget of interesting content from what was no doubt a dry and overly verbose dissertation, and rather than sending me a computer-generated rejection letter, encouraged me to take a few more kicks at the can.

I am humbled by the generosity demonstrated by the women who donated their time and whose stories appear in this book. All of them were willing to discuss their experiences with a near-complete stranger for free, and to provide wonderful comments, criticisms and insights. Without them, this research would not have been possible, and it is my wish to "give back" to them by recording some of their histories in this document. The leaders of DigitalEve welcomed me into intimate gatherings, where I was allowed to observe private discussions of sensitive organizational matters. Anna Gonowon spoke at length with me and provided some excellent insights drawn from her experiences both in feminist scholarship and within the information technology industry. Christine Pilkington, former head of DigitalEve Toronto, was likewise forthcoming. Heather Finley, Allison Fraser and Kathleen Webb, all current or former DigitalEve executives, took time out of their packed schedules and met with me late one evening after work for several hours in order to let me ask pointed questions about organizational governance. Jennifer Evans, a former head of DigitalEve Canada, was truly amazing in her commitment to my PhD research and deserves special mention. She facilitated meetings, made introductions, enabled me to access the inner workings of the organization and, most importantly, was unflinchingly kind, honest and helpful throughout.

I have also benefited greatly from supportive family and friends. My father told me always to cite my sources properly, and my mother told me never to eat the potato salad in restaurants; thus far I have been relatively free of both plagiarism charges and food poisoning. Sandeep De reminded me of the importance of using a materialist paradigm, helping me to remember that what is done is as significant as what is said. Jeremy Adirim delighted in pulling back the curtain of the IT industry to reveal many of its conceits and fictions. Sean Davidson unleashed his editorial zeal upon my inappropriate use of obscure conventions of punctuation. Rob MacDougall, the hippest historian that Harvard has ever produced, somehow managed to combine *Simpsons'* quotes, deliciously puerile humour and brilliant, incisive feedback.

And finally, I would like to thank my wonderful husband Chris for his unfailingly cheerful support of this research in numerous ways, from scouring electronics shops for the very best deal on a mini-tape recorder and a swanky computer, to explaining the intricacies of Excel spreadsheets, and above all for providing the emotional sustenance necessary to complete first a PhD and then a book. He is, and will always be, the light of my life.

Knowledge is produced as a result of relationships, not by one person doing intellectual labour in a vacuum. I am indebted to all who participated in the generation of this work.

INTRODUCTION

WOMEN'S WORK IN INFORMATION TECHNOLOGY

Every technology is both a burden and a blessing ...

— Neil Postman, *Technopoly: The Surrender of Culture to Technology*

THE 1990S SEEMED TO BE a time of infinite possibilities for the information technology (IT)[1] industry in North America. Growth was exponential, rapid and apparently limitless. From 1992 to 1999, the expansion of the IT sector was three times that of the total Canadian economy.[2] During this time, women moved into IT professions, attracted by new kinds of jobs and the promise of lucrative, creative, interesting work. According to breathless futurists of the period, we would all speed down the information highway, riding our converged media, doing a booming business in a "friction-free economy," where "demand effortlessly follows production."[3] Wealth would be easily and infinitely generated in an economic parthenogenesis. The screech of the voluminous bandwidth on our souped-up e-vehicles would merge with champagne corks popping to provide the soundtrack of our lives, as our little tech startups mushroomed into multimillion-dollar enterprises. Work would be plentiful and well compensated and performed either in the comfort of our homes or in the paperless office. In our many hours of leisure time as IT consumers, we would shop on-line, download various media of our choice, buy nifty gadgets, engage in mind-expanding cybersex and enjoy access to limitless information as part of the global cybervillage. Gender, race, age and other troublesome markers of identity would become meaningless artifacts of a shallower, bygone age.[4] It was a vision of the technological future that rivalled the atomic 1950s in its starry-eyed optimism and firm foundation in consumer capitalism. According to mainstream publications of the IT industry such as *Wired* and *Fast Company*, IT work and culture were hotter than a spastic microchip. "Visions of the future," an advertisement in *Wired* magazine breathes heavily, "come from a place where we have the power to do anything."[5]

Other observers were not quite so optimistic. Some hand-wringing prophets of doom took the opposite tack, in varying degrees of nervous nihilism. They pointed to the vast disparities in global access to information and communication.[6] They worried that we would evolve from tool-using to tool-oppressed *Homo sapiens*. Human tasks, they suggested, would be fragmented into partial, mundane and repetitive pieces, while computers, perhaps delighting in their intellectual prowess, would run resplendent algorithms in the service of capitalism. Decent work would die an indecent death.[7] As for women, sources debated what negative consequences this technological dystopia held for them. Damsels in distress might find themselves in need of defense on the electronic frontier as Internet pornographers, on-line rapists and cyberstalkers massed at their digital and physical boundaries. Mothers-to-be might find their pregnant bodies under surveillance, their fetuses becoming celebrities in ultrasonic reality television.[8] Women might undergo the controlling regimes of workplace monitoring, or perhaps real women might be eliminated altogether with the emergence of virtual, computer-animated sex kittens.

Each of these visions is correct in its own way. Each identifies different kinds of social anxieties and hopes and the ways in which information technologies might mediate them. These models of the future serve as a warning against ahistorical exuberance over technoscientific progress, as well as a paean to human invention and creativity. But in the early twenty-first century, neither is true while at the same time both are true. Rather, as we clay-footed humans are wont to do, we muddle through, stumbling and weaving over a middle ground. The mundane minutiae of people's daily experiences with information technologies have smoothed the cutting edge of the "information revolution." At the same time, the banality of these technologies can conceal their potential to enable dramatic changes in work practices.

The history of the so-called information revolution in the 1990s is a contradictory one. Change has been dramatic and rapid and has introduced new possibilities and challenges. During this decade, social and economic inequalities were both alleviated and exacerbated. In 1991, when I began my undergraduate studies, I was one of the very few students in my residence who brought along a computer. Most of the other students used electronic typewriters. Now, there are Canadian universities who dispense laptops to their students as a routine matter. Those of us who had the privilege of learning how to use computers learned to make our communication and representations abstract, dispersed into networks and webs rather than lines. Users in fields like arts, social sciences and humanities discovered the vast potential of new kinds of software for creating new kinds of research and projects.

On the one hand, the emergence of vast communications networks enabled new forms of work, new types of allegiances, new dialogues between individuals and groups who might previously have been geographically or temporally separated, and new ways to trade information. Groups of workers who were used to being socially marginalized as nerds, geeks and dweebs, and mocked in movies such as *Revenge of the Nerds,* suddenly acquired hipness and lucrative jobs. On the other hand, these communications networks were rapidly appropriated into the service of globalized, transnational corporations. Certain types of work disappeared, either to reappear in another part of the world where labour was cheaper or to be erased altogether from the social hard drive. The enthusiasm of North Americans for information technologies required the labour and raw materials of workers in other spatial locales and economic strata, and the world continued to split along fault lines of producers versus consumers.

Employment in this new field of information and communications technologies was strong, increasing consistently throughout the 1990s after rebounding from the early 1990s recession.[9] Many workers benefited from the Y2K paranoia, although many were considered self-employed contractors and were let go as soon as the threat of a technological apocalypse subsided. Though women continued to be under-represented in the field of IT, particularly at senior levels, they nevertheless managed to comprise about a third of the workers in IT industries. However, the decline that began in January 2001 was relatively sudden and precipitous. Particularly hard hit was the communication end of the information and communications sector, which in its painfully wounded convulsions managed to take down other industry sectors with it. There were massive cutbacks, layoffs and bankruptcies across the board, from manufacturing to trade to service-based IT companies. Though startups and small businesses had traditionally been ephemeral, their debut and finales occurred on a much larger scale, and often with greater rapidity. Large IT corporations such as Cisco, Hewlett-Packard, Nortel, WorldCom, IBM and many others also took a solid thwack to the economic solar plexus, and many collapsed in paroxysms.[10] In the massive layoffs of workers that resulted, women suffered disproportionately. Their job loss far outpaced that of men.[11]

In our more sober era, what might have been viewed as a radically new way of doing business in many IT companies in the 1990s could now be generously termed "accounting irregularities."[12] Now, businesses that focused on gobbling and spewing out an information technology infrastructure might remind cynical investors of the tulip bulb craze that ruined Dutch speculators in the early 1600s.[13] Salaries have descended from the stratosphere. Cubicles are emptying, and the surviving employees are beginning to ask pointed questions about benefits, work hours and the viability of stock options.[14] Some have even joined fledgling unions.[15] While many people, including myself, enjoy technologically facilitated homework, some workers with pagers and cell phones are wondering whether eliminating physical workplace boundaries has also resulted in eliminating practical, pragmatic work boundaries such as a job site that one can leave at the end of the workday. Canadian industries and governments lament the shortage of skilled workers while "outsourcing" jobs to other companies, regions or even countries where labour is cheaper. Worker stress is at an all-time high, and bodies are breaking down under the strain of repetitive, technologically constrained tasks.

And what of women who currently work with information technologies? Both models of the future concerned women on some level, as subjects, objects, actors and victims. What work are women now doing in the IT field in Canada? Where are they most likely to be found, and why? Has the expansion of the information technology field represented a significant shift in how work is experienced and performed and, if so, what form has this shift taken? What opportunities have women found, and what obstacles?

These questions are deceptively simple. It can be difficult to get a clear picture of women's work in the IT field. For example, one of the most common places to look is at enrollments in technical education. We can tell with a brief glance at educational statistics that women's enrollment in post-secondary-level computer science has stagnated or declined, which is an interesting fact in itself, but a computer science degree is most certainly not a prerequisite to be an IT worker.[16] Statistics about the gender breakdown of the labour force can be hard to interpret, since IT work cuts across many industrial and occupational categories. Some IT workers are hidden at home, doing contract, freelance, part-time or other forms of precarious work. Some IT workers do technical tasks as part of their daily routine, but their official job title is not considered a technical one. Information technologies and changes in work practices have created many new types of jobs and reconfigured old ones. What I call hybrid jobs — jobs that combine various backgrounds, disciplines and types of tasks — have emerged. IT work, then, is a category of convenience rather than precision, which suggests a loose grouping of work types that can vary by actual tasks performed, by work practices and by the relationship between employer and employee.

What is officially defined as IT work is in itself a political issue. IT work, as an idea, is highly valued, often near-mystical. IT's origins in the masculine world of the military, computer science labs and technical institutes still lend it a character of unassailable worth. Non-IT workers — I use this term advisedly, since all forms of work have been changed by technological developments — may not be sure what, exactly, IT workers do in a day. They often cannot understand it even if the IT worker is generous enough to explain it (and this generosity is a rare pleasure, since many computer-proficient people look down on "newbies" with open disdain). Yet non-IT workers may not fault or question the IT workers, because they may assume that the field is simply incomprehensible to the uninitiated. And, indeed, it can be: the common language of this work resembles medieval Kabbalistic incantations in its cryptic syntax, invented words, numerological symbols and peculiar verbal rituals. Let us say, for example, that I ask my sysadmin (systems administrator) how she resuscitated my e-mail after it choked on a piece of rather bony code. She might respond, "Well Krista, I telnet'ed to 110 and logged into the POP3 server manually. The LIST showed me the large message, so i RETR'ed it, captured it in my term buffer, and DELE'd it so that you could POP without stupid Outlook croaking every time its parsing engine hits non-standard JavaScript in text/html." If she were like many sysadmins, she might serve up this plate of gibberish with a side order of implied contempt for choosing "stupid Outlook." At this point, I would probably scurry back to my office and the odious Outlook, too afraid to make further inquiries.

Shamed by their ignorance, made painfully aware of their ineptitude and on the receiving end of messages about the importance of a "knowledge economy," non-IT workers may content themselves with the assertion that IT work is complicated, challenging and, above all, *crucially important.* In our current social and historical context, there is a tendency, facilitated by consumer capitalism, to suggest that information technology and "innovation" in general should be an inherently valuable commodity. For example, a document produced by Human Resources Development Canada, with the grandiose title *Achieving Excellence: Investing in People, Knowledge, and Opportunity,* insists that companies "see innovation as *the* way to grow profit margins and increase productivity." If innovation powered by information technologies is valuable, then people who use technology must also be inherently valuable. Conversely, perhaps only people who are seen through the lens of social norms to have inherent value should be permitted to control technology. In any case, participation is a group obligation. "If we are to succeed," the document intones, "innovation must be everybody's business."[17] Information technology has become clearly linked to social progress and social responsibility. A good citizen is a technical citizen. Good work is knowledge work.

Despite their current ideological omnipresence, concepts such as innovation, knowledge work and information technology tend to be symbolically bloated notions that are divorced from specific, often contradictory, experiences. What is designated officially as "technical enough" to merit classification as knowledge work reflects social perceptions of how IT labour should be judged. Inherent in these social perceptions are also assumptions about the value of who does this work. This value is implicitly or even explicitly gendered. For example, there is some tension over how to understand call-centre work. In terms of task performance, call-centre work could certainly be viewed as IT work, merging as it does various types of information and communications technologies.[18] It is sometimes included in hype about the "high-tech" economy, such as in recent Maritime initiatives.[19] On the other hand, call-centre work may also be hidden and viewed as unskilled labour. The bulk of call-centre work is done overwhelmingly by women and is performed under the umbrella of industries such as banking and finance, rather than "high-tech" industries per se. It may be seen as an extension of women's clerical work, simply another way to be a bank teller. If women do work in IT industries such as telecommunications, their role may also be hidden because the work, such as telephone operation, is considered menial or non-technical. Since the 1990s, this work has slipped quietly away, jobs eliminated by layoffs, contracting out and the technology itself. Is it accidental that much of this devalued IT work, which is methodologically concealed and technologically eliminated, is women's work? Perhaps. Perhaps not.

Despite significant categorical changes in the nature of certain kinds of work, many new opportunities for women in the field and the diversity of women's work in IT, women's experiences of gender stratification in the labour force and workplace have remained remarkably consistent. Women, as a group, continue to be paid less and do more unpaid work than men, and this continues to affect their choice of work.[20] Although many female IT workers love their jobs and the work they do, they continue to experience workplace harassment and both overt and covert discrimination based on their gender, race/ethnicity, age, class and sexuality. Their work continues to be undervalued and their home-based paid work is often seen as an extension of their role as homemaker. They continue to be a small minority in a male-dominated field,[21] and even when they work for IT companies or in technical industries, they are more likely to be doing female-typed jobs. They are barely present in technical fields in universities, and if they do enroll, they soon drop out. They often find their way into IT accidentally, as a result of a non-linear career path, rather than as a deliberate choice.[22] Even more worrying is that the small gains women have made in the latter half of the twentieth century may be eroding in the face of larger labour market shifts that render jobs more tenuous. Increasingly, workers are getting by (or not) with a collage of contingent jobs, including part-time and contract work, as well as solo self-employment.[23] When speaking of overall socio-economic conditions, people often say that a rising tide lifts all boats. It's clear that a persuasively pulling undertow can also suck them down.

How can we explain these persistent and problematic trends? IT was supposed to free us from conventional markers of identity and level the playing field for everyone. Rational, traditional theories of work choice would tell us that women's under-representation in the IT field is because women are simply uninterested in participating in the IT field, whether by choice or design. Many well-meaning initiatives still struggle to interest girls in technology, and often imply that there is some deficit in females, some attitudinal tic or experiential absence that simply needs to be corrected. Earnest psychological literature suggests that women just can't hack the tech, cognitively speaking, perhaps because of the size of their hypothalamus or their relative levels of endogeneous testosterone, while popular evolutionary theories might even go so far as to construct elaborate links between the imagined behaviour of cave-dwelling *Homo sapiens* and the present behaviour of cubicle-dwelling *Femina sapiens technologica.*

But simplistic explanations of women's inherent technological deficit and dislike do not sit well with me. Metaphorically speaking, it's like trying to ram a parallel connector into a USB port. I can probably make it fit with a sufficient application of force, but it probably isn't going to work all that well. I have heard too many women speak with passion about their work in IT, about the thrill in solving a difficult problem, about the ease with which difficult tasks can be completed, about the world of possibilities which the technology has created for them, to believe such simple dismissals of women's relationship with IT. "I am in love with computers," says one woman to me. Another woman tells me about the intuitive, excited feeling "in her gut" that she gets when finding a solution to a technical problem. A third woman, not in the least fearful of technological objects, gleefully performs "percussive maintenance" on her hard drive by giving it a whack when it misbehaves, to jiggle the hard drive back in (I confess to doing this myself whenever the computer fan rattles in the case). In a survey sponsored by Statistics Canada, over half the respondents report that the introduction of technology has made their work more interesting.[24] Another report notes that "women experience working in the new economy with an exhilarating sense of achievement, impact, satisfaction, and opportunity for creative freedom they didn't have before."[25] And thus, because of these types of narratives, I am compelled to look deeper to explain women's experiences in the IT workplace. What I find is that IT work for women is complex and contradictory, neither wholly negative nor wholly positive. IT work can constrain and liberate, restrict and empower women. Women's situation in IT both reflects and challenges norms of women's role in the labour force.

ANALYTICAL APPROACHES TO UNDERSTANDING WOMEN'S WORK IN IT

Perhaps the most insurmountable theoretical gap in our traditional understanding of women and work is the lack of attention to complex social positions of race, gender and other intersecting structural relations as *fundamental* (not tangential) to understanding women's work. In other words, women do not go to work and then have gender, race or class "happen" to them; rather, their lived reality of these things is entwined in their work experience before they even set foot in a workplace; it guides their choices and is ever-present in daily life. They may not think about these things much of the time, especially when they are in a context where other people look and act like them. Yet occasional reminders intrude: the worker might wonder why she is always asked to fetch the coffee, why her paycheque seems leaner compared to her peers, why she can't quite seem to get ahead in a particular company, why she feels uncomfortable when she is exhorted to be a team player or "one of the boys," why the office hierarchy develops in certain ways, why co-workers in a meeting seem unable to hear her speak. But markers of group identity, such as gender, and the social value assigned to them, extend their influence beyond individual affronts or the day-to-day nuisances of the workplace. Traditionally, the capitalist labour market has depended on divisions between groups of people — immigrants versus nationals, non-unionized versus unionized, skilled tradespeople versus unskilled day labourers — to fuel the machinery of production, as well as to prevent worker solidarity. Certain types of people can be counted on to fill certain types of jobs, and strata of pay, status, working conditions and job availability are maintained with surprising reliability.

Talking about gender and work can be a delicate and difficult project in a world which often insists that people succeed or fail entirely on their own individual merits, and which sees the concept of gender as a generalized "women's problem" that has recently come into vogue. Recently, I witnessed a female academic colleague arguing politely with a male researcher, a man who studied economic statistics. He insisted on saying that she studied women. She corrected him and said that she studied gender, not women per se, but rather the relationships in society between men and women. He rolled his eyes. "Yes," he said with affected patience, "but *everyone* knows that 'gender' means 'women.'" He seemed not to notice that he himself had a gender. Issues of gender stratification in the workplace are often not seen as relevant when speaking of men, since women are seen to be the only ones to "have gender." But gender is not just a thing that each individual has; it is an analytic concept that allows us to understand the relationships between people. Gender is something that organizes the world of work, and its effects are differently felt by men and women. Men are not immune to a gender system either. For example, research on male hackers and engineers demonstrates that complex cultures of masculinity, with attendant class and race/ethnicity dimensions, emerge from certain technical practices.[26] Norms, while diverse and at times contradictory, are nevertheless grounded upon certain shared values about what it means to be a technical male, and are reinforced enthusiastically by group members through various practices such as sharing (or withholding) knowledge and granting community access through hazing or gatekeeping rituals. But there is, of course, more to this than which washroom one uses. Each woman has other identities that mark her visibly and invisibly: race/ethnicity, class, age, sexuality, ability. White women are not seen as "having race," and middle-class women's position in a class hierarchy is invisible. But just as with gender, these identities form relationships in society. There are vast differences among women workers, and often deep divisions of power and privilege, which is why it is difficult and dangerous to generalize based on gender alone.

Pointing out differences among women does not mean that this book is about a cacophony of identities. Women are not "just" raced or classed or subject to wearing a variety of "identity hats" in isolation from one another. Identities must be understood as relational instead of categorical, context-specific instead of predetermined, multiple instead of singular and dynamic instead of static. Our selves are inseparable and indivisible. Our identities are messy and bleed into one another. To whom (or what) do we claim primary allegiance? What team do we cheer for? Are we moms, daughters, sisters, partners, workers or citizens first? Are we primarily defined by our race, our ethnicity, our first language, our gender, our age, our choice of partner, our geographical location, our abilities? We cannot answer these kinds of questions because we cannot fragment ourselves this way. We are all of these things, and more. Moreover, these things have a life outside of us, and they are more than individual characteristics which we possess. For example, being a mother is a significant, defining experience which one person may have, but the way that motherhood is understood and organized in our culture supersedes that one person.

Though social factors are important, this isn't intended to be a causative or prescriptive model. In simpler terms, who you are doesn't necessarily determine what you do. While structural relations undoubtedly shape women's work experiences, they do not entirely limit them. Being Person A doesn't inevitably lead to Experience B. Rather, I would like the concept of multiple structural social relations to be a filter through which women's work experiences are read and against which theories of labour are tested. How does being female in a workforce, which we know is differentiated and stratified by gender, affect a woman's work identity and experiences? What tensions emerge? What contradictions? How does she fit and yet also defy the norms? How does she struggle with and also reproduce relations of inequality?

After speaking to women about their work and listening to their stories, I believe that women's awareness of their own situation allows them to make choices from the options they feel are available to them. As they progress through their work "life cycle," they learn what is expected of them, what obstacles they are likely to encounter and what strategies they can use to survive. Given similar circumstances, women can make different choices. Or, their journeys may be unique, but their destination common. I apologize to anyone who is reading this and looking for advice on how to write *Mars and Venus in the IT Workplace,* but I cannot say that *all women do X* or *all women feel Y.* Such a simple, binary theory feels cozy and comfortable, as it provides the basis for much pop psychology and mainstream analyses of women's choices. How easy it would be to divide the world into pink and blue, Barbie and G.I. Joe, sugar-and-spice versus snails-and-puppy-dog-tails and be done with it! But we must not. We cannot. Women have told me stories that both defy and reproduce gender norms. Few of us have a straightforward work narrative; most of us experience twists, turns, tensions and surprises as we move through the labour force. If the theory doesn't fit the evidence, then one has to change the theory, and Mars and Venus aren't going to cut it.

Yet, as I examine qualitative and quantitative evidence, data of all kinds, it is irrefutable that there are clear, definable, measurable trends that shape women's experiences in the labour force and that serve as frameworks that help to organize women's lives. How, then, do I pick my way through this thorny path? How can I theorize women's experiences that, on the one hand, are unique to each one of them and, on the other hand, reflect persistent themes? I prefer to think about this as a relationship between elements that are constantly in flux, always changing and reconfiguring themselves. Gender (and race, and age, and a host of other things) is not, then, merely a characteristic or capacity of individuals but a way to enter into a relationship with the world, and a way in which the world is organized. This kind of relational model necessitates articulating contradictions and tensions, as well as critiquing normative standards and facile categories of "difference." One cannot discuss women's experiences in the labour force without also examining their intersecting identities and multiple structural oppressions. We must think of women in the work force as both individuals with complex identities *and* people who are participants in a series of relationships that we label as gender, race, class and age, among others. These relationships are not completely prescriptive, but are nevertheless fundamental to our experiences of ourselves and the world around us.

It is essential to situate this research in material, or "real" and lived, practices of women's work, because it produces a grounded analysis that examines what women are actually doing and experiencing. While it necessitates concrete observations of many basic things, it also allows for a theoretical perspective that can explore individual contradictions and tensions as well as structural and systemic trends. This perspective could be called a *materialist* one because it is concerned with these material conditions (not to be confused with another sort of materialist perspective most famously endorsed by Madonna). A materialist type of theory is helpful in studying the workplace and labour market, because we are often confronted in our culture with information about what women, as a group, "really" want or "really" think about, and this information appears to contrast with what actual women do in their daily lives. This tension intrigues me, so in this book I ask about what women are doing from day to day: How do they put food on the table? Do they like their jobs? What do they do all day (or all night)? How is each woman's work experience individual to her and at the same time resonant with the experiences of others? One theoretical question that frames this book is: *What are the material conditions of women's work in information technology?*

Many materialist analyses rely on empirical evaluations of women's experiences and things that can be quantified. I will rely on both quantitative and qualitative assessments of women's work experiences, since experiences include ideologies and feelings as well as concrete material conditions. I will present some numbers here and there, but will try to bring those numbers to life by wrapping them in a story, a history, an idea or an image. There are often fascinating contradictions between what is apparently experienced and how that experience is understood. In the space between experience and understanding may lie an intriguing nugget of information for us about how people make sense of their world. Moreover, instead of assuming some kind of "natural," direct, unmediated experience, I will begin to theorize about relations of power and how they filter experiences, both ideological and concrete, as well as the meaning that is made from those experiences. In this model, people are agents who make choices from the options that they feel are available, and these options are a combination of what is "actually" present (i.e., concrete material conditions) and what is "seen to be" present (i.e., perceptions of risk and reward for each choice). Thus, people's experiences do not just "happen" to them; rather, they are both mediated through material and ideological power relations in their occurrence and are interpreted through a framework that is strategic and subjective.

Using this type of approach to examine women's work in the IT field, it is immediately apparent that, as I have noted, while we can observe particular trends in women's work, we can also see a great deal of diversity. There is not just one story of women's work in IT; there are as many narratives as there are women, and their multiplicity shows us that a vast array of jobs, practices, work identities and experiences are possible. Women in IT are doing every job at every level of seniority and skill. Unlike much of the technology that has come before it, IT is an extremely versatile tool for performing work and inventing new types of work. In addition, women who perform the same type of work may understand and experience it differently. Thus, just as traditional theories about women's work experiences and choices are inadequate for understanding a labour market which continues to be stratified by class, race and other structural dynamics, theories of "women and technology" that propose a single, gendered type of technology use and interaction likewise do not tell us the full story about women's work in the IT field.

In the course of writing this book, I used a variety of methods to gather information and tell the stories of Canadian women working in IT. I used the voices of government and business in the official documents that they produce, as well as the statistical data they provide. I used theoretical literature on women, technology and work. I used surveys and information from industry associations and services. I used data collected through fieldwork as well as over sixty in-depth interviews with women working in or studying to work in IT. The women I interviewed shared only three common traits: a gender identity as women, an interest in technology and past, current or desired work experience in IT. In all other areas of demographics — personal identity, language, country of origin, sexuality, age, expertise, education, income level — they varied widely. This fact of variety is significant in and of itself. It tells us again that we cannot create one single narrative of women in IT. In the text, I move from theory to experience, meandering through life histories, but from this collage of research methods, patterns emerge. One of the ways of checking validity in qualitative interview research is to look for motifs that appear repeatedly in similar ways. Repetition of familiar refrains gives coherency to what, at first, appears to be a mishmash of stories. Women featured in this book are, in general, concerned with the same things: the quality and value of their working life, gaining technical knowledge and expertise, struggling to define their role in a generally youth-oriented, male-dominated field (or, in the case of many women, to carve out a technical role in a traditionally female field) and trying to juggle competing responsibilities and meet diverse expectations. How they approach these challenges and the meaning they make of them are, however, unique. In some ways they share things that are common to working women in general. In other ways their situation is characteristic of IT work. It is, in part, this contradictory quality to their stories that intrigues me.

While this book is meant to be a snapshot of the IT labour force at a particular point in time, it has limitations. Despite the advantages of virtual communication, the constraints of geography are compelling, and the majority of my interviews are drawn from women in the Toronto area. For one thing, Ontario is home to 44 percent of the professional, scientific and technical services industries, and to 41 percent of the information, culture and recreation sector industries in Canada. The province also hosts 40 percent of all the Canadian occupations in the natural and applied sciences. Toronto, unlike many other regions in Canada, has experienced the IT boom and bust in a way that is specific to a big city with a comprehensive infrastructure (including high-speed Internet access), a large and diverse population and a well-established "knowledge industry." There are both pockets of low-paid work such as call centres as well as substantial corporations that provide employment and some degree of financial stability. There are many public and private institutions and organizations that enable one to acquire skills, training and contacts. Nevertheless, while there are particularities to Toronto experiences, the patterns I identify are present in all regions of Canada, albeit in different forms, and they are situated in an even bigger picture of global shifts in labour practice. For example, in my discussion about call centres, while I discuss call centres in Toronto that take advantage of a labour pool that is marginalized, I also point out that these operations often locate themselves in economically depressed regions of Canada; I suggest that this is not accidental, but rather takes advantage of a labour force that is similarly marginalized and desperate for work as a result of the loss of local and indigenous industries. Jokes about Toronto fancying itself the centre of the universe aside, it in fact serves as a useful microcosm for examining the relationships, linkages and tensions between high- and low-end technological work, between technical haves and have-nots, between local, national and transnational labour arrangements.

In this book, I also focus primarily on women working professionally or semi-professionally in IT, rather than on women doing low-end work such as data entry or on women working with manufacturing technology. This is a deliberate choice of focus on my part, but not a malicious exclusion. For one thing, the richness and complexity of the experience of low-paid technologically facilitated work cannot be adequately addressed in this text, and to pretend that I do so here would be a disservice. However, I think it worthwhile to indicate my awareness of this work, and I try to do so throughout the book. But more importantly, my central aim is to target the more privileged forms of work that have remained more apparently equitable and interrogate this veneer of formal equality. When we hear about the lucrative possibilities of IT work, it is clear that low-status IT tasks are not included. But a critical, feminist inquiry into the forms of work that *are* suggested as positive and empowering for women is worth doing, if only to closely examine the promises made.

In this book, I suggest that, in fact, the low-status work exists in a dynamic relationship with high-status work, and that one requires the other. For example, the low-paid subcontracted labour of salespeople who sell high-speed Internet connections door-to-door helps to create a communications infrastructure that then sustains the professionals who maintain and develop it.

Moreover, the high-status work — in that it is increasingly "credentialized" (subject to formal qualifications rather than informal, experiential knowledge), subject to managerial controls, deskilled through automation of the labour process, precarious and scarce — is starting to resemble low-status work more and more. The mythic days of the tech cowboys who, fuelled by caffeine and loud music, revelled in their unconventional workstyle while watching their bank accounts swell, are no more (if they ever really were). In their place are tired, stressed, often-downsized workers who struggle to find decent work with job security, benefits and support for family life. While they have been told that the hard skills of computer mastery and certifications for which they have worked hard to acquire are what matter, they are also told in interviews that their "soft skills" of caring and communication, which computer science neglects but which customers expect, are what really counts. They produce less and less, and serve more and more. Thus, from one angle, distinguishing between a call-centre worker and a software-services employee is merely a matter of scale and location, rather than one of substantive difference.

Who are the IT workers who are included in this book?[27] They work in a variety of areas: health care, public service, network security, software development, graphic design, telecommunications, education. What they have in common is some kind of information technology task performance as a primary part of their work practice and work identity. This excludes people who use computers only to type letters or to make spreadsheets. Such a task performance can occur at the level of hardware, software, creation, development or use of information technologies. As IT is more fully integrated into more workplaces, the boundary between who is an IT worker and who is not becomes increasingly fuzzy. "IT worker" is often a category of convenience, which can signify a particular approach to the objects and culture, a particular skill set, a particular orientation or position within an organization or a particular identity. In general terms, however, IT workers tend to operate in the following capacities: electronic data management and storage, IT infrastructure, networks and network security, software development, communications technologies such as voice and wireless, Web work such as Web design or graphic art, electronic commerce, intranets/extranets and communications applications such as on-line marketing.

More importantly, I have included women by their *self-identification* as IT workers. This fact is interesting in and of itself, for it points to one of the contradictory truisms of IT work: as it becomes increasingly formally defined, it also becomes increasingly resistant to categorization. More than twenty years ago, when the 1980 Standard Occupational Classification was implemented by Statistics Canada, there were only a handful of occupations linked to IT. Now, labour analysts scramble to keep pace with about two dozen general categories of occupations, adding more levels and deepening the complexity of existing ones.

So, again we return to the central questions that frame this book: At this moment in Canada, what IT work are women doing? How? Why? Where? How are their experiences mediated through both structural and individual factors? And where might they go from here? Throughout the rest of the book, I explore these questions in the context of factors that shape work experiences. Each of the following chapters focuses on a central contradiction or area of tension I wish to highlight. Chapter 1 identifies some themes that frame an investigation of women's work, and suggests that women's work in IT is a historical product with a material base in social relations, rather than a truly new form of work that radically reconfigures work arrangements and experiences. Chapter 2 takes a closer look at an explanatory theory of women's work choices and sets it alongside current discourses about the role of skill in IT work. In this chapter, I argue that the notion of skill is itself gendered, and that reducing women's IT work experiences to skills alone erases structural relations of gender, race, class and other determinants that mediate these experiences and reduces each worker's experience to simplistic individual factors.

Chapter 3 recalls some of the great promises of the IT industry of the past, such as the possibility of unlimited well-paid jobs, and situates them in the current realities of the industry for women. Chapter 4 revisits the theme of new versus old introduced in chapter 2, to show that while IT work can be lucrative, creative, exciting and liberating for women, it also replicates the hierarchical social relations of the labour market. Chapter 5 concludes by looking ahead, renewing the spirit of what I would call "informed opportunity" or "critical optimism," which I believe should characterize women's approach to the IT field. I show ways in which negative patterns are being disrupted and ways of doing business are being challenged, and suggest possible directions for the future.

CHAPTER 1

Women's IT Work in Context

> Life is not determined by consciousness, but consciousness by life.
>
> — Karl Marx, *The German Ideology*

> The liberation of women — and all human beings — depends on understanding that work is essential to our development as individuals and on creating new places in our lives for our work.
>
> — Nancy Hartsock, "Staying Alive"

BRENDA'S HOUSE is one in a series of modest little boxes with tiny neat lawns and gardens in a working-class neighbourhood. Brenda has carved her workspace out of a corner of her small dining area, and the room feels like a cozy nest of papers and computer paraphernalia. Brenda herself is a hearty woman with a generous laugh and a maternal air. She tells me that she worked as a waitress, and I imagine her as the type of jovial server who calls everyone "Hon" and cheers up the regulars at the truck stop with extra refills of coffee and slightly off-colour jokes. She wants to build on her interest in people and technology, so she has started her own business assisting other entrepreneurs to develop electronic commerce startups.

She honks out a guffaw when I ask her about her education. "Oh yeah, I went to college for about ten minutes. I really excelled at euchre." Fiercely independent, clever but not interested in conventional avenues of education, Brenda travelled across the country in search of adventure. She settled for a while in the tourism industry, but felt uninspired by the lack of options for career advancement and personal interest. Looking ahead, she said, "Either you own the hotel or the restaurant, you come by and pick up a cheque and sign some forms. Or, you're the lowliest worker. You don't care if the place burns down. You get your $8.50 an hour or whatever. What I saw down the road was middle management, and I said no, I'm not going to spend the next ten years working towards that. Why?" Brenda is keenly interested in finding value and purpose in her work. "How do you know that you really like what you're doing? Would you be willing to get up and do it on a cold February morning for free?" Despite her lack of formal education, she is unwilling to settle for a career that stifles her soul.

Work as Central to Life

For most of us, work and activity are central to our experience of the world. At parties, to make small talk, we ask one another what we do for a living. Our bodies bear the scars of our labours, paid and unpaid: hands reddened, sliced and burned from cooking; wrists tingling and numb from keyboarding; back aching from lifting; feet swollen from standing; waistline expanding from the takeout food we spill on ourselves and our steering wheels as we attempt to navigate gustatory and intellectual functions on our way to a client meeting. We often identify ourselves by our jobs, though this designation tends to be reserved for professionals: I'm a lawyer, I'm a doctor, I'm a teacher. Not so many of us will say: I'm a shelf stocker, I'm a package mover, I scrub floors, I am paid minimum wage to smile at strangers. Nevertheless, our work and activity shape much of who we are and who we perceive ourselves to be.

Throughout this book I use the term "social location" to refer to the complex interactions between our gender, race/ethnicity, age, ability, sexuality and socio-economic class. In a literal sense, it means where and how we are positioned in society. Social location is foundational to our experiences of the world. It affects things such as the way we interact with others, which communities we feel a part of and how we inhabit social and physical spaces. Social location includes the individual characteristics and social signifiers, such as skin colour or speech patterns, that we possess, or roles that we play in our everyday lives. However, I also use the term to express relationships: between us and others, between ideas and between social structures. Social location permeates our experiences, popping up when we least expect it to, and reminds us acutely of the relationships of power that continue to exist in our society. What shapes our experience of work, then, is a variety of these political, economic and social relationships. These relationships may be large scale, and they may be asymmetrically weighted based on which individuals or what institutions tend to have power and privileges. These relationships may be small scale: they may be communities or workplace dynamics. Whatever their scale or scope, these relationships always have a context and are constantly in flux. They also have a history.

In this case, I am beginning this examination with double vision: looking generally at women's work and specifically at women's IT work. I will point out patterns that are endemic to women as a group but also point out factors that are specific to the evolving IT field. Women's experiences with IT work are diverse and are influenced by power relations. While there are clear connections between elements such as gendered patterns of labour and the male-dominated, masculinist culture of some IT workplaces, I do not suggest that this relationship determines the experiences of all women. Rather, women's IT work experiences are a product of both structural and individual factors.

It is easy to suggest that information technology itself bears sole responsibility for changes in the nature of work. After all, the deployment of global webs of information and communications technologies have created networks of data that span the globe, refuting the limits of time, place and speed of discourse. The scale and scope of their reach have altered nearly all forms of work. But IT is not the only agent of this change. A computer in Canada cannot leap from its cardboard box, put its own parts together and link itself to an electronic mate in Malaysia or Berlin. People imagine and build information technologies. They distribute, implement and connect them. Most importantly, it is people and social institutions, acting with specific interests, that determine how and why these technologies will be used. Both technology and work shape, and are shaped by, a complex web of social, economic and political relationships.

Along with being complicated, these relationships may also be contradictory. People's experiences with information technologies, and their experiences as workers, are not straightforward but often represent persisting tensions and struggles over meanings and practices. Indeed, the very features that make these things appealing may also make them irritating. For example, IT transcends time and space. This may be an advantage in one context, perhaps when a researcher needs to retrieve an article from a library at 3 a.m., but a disadvantage in another, such as when an employer gives an employee a pager with the understanding that the employee is now always on-call. Information technologies both enable and constrain women's work in a variety of contexts. The women I spoke to would all say that in general, information technologies have improved the quality of their working lives. IT has allowed them to perform new kinds of tasks and do old tasks more efficiently. They enjoy their jobs and are pleased with their technical aptitude. However, the practice of IT work is not gender-blind. Computers do not necessarily change familiar patterns of behaviour, and the elements of IT work that can provide opportunities can also reinforce traditional disadvantages for women.

Patterns in Women's Work

While it is inarguable that women have made significant gains in the arena of paid labour in North America, especially since the Second World War, there are some fundamental trends that remain unaltered. Despite an increase in female participation in the labour force, women remain concentrated in particular sectors of the economy and experience a disparate load of unpaid and undervalued work, industrial and occupational segregation and pay inequity.[1] One striking feature of the historical record of women and work is how constant these themes are. Despite many increases in women's labour force participation, skill level, access to education and advocacy, there remain fundamental divisions in the organization of paid and unpaid work. These divisions, in part, depend on social location. In general, groups of people that were doing certain jobs decades ago are still likely to be doing them today. For example, unpaid and low-paid caregivers still tend to be female. Unionized tradespeople still tend to be male. The content of the work might change over time, but the form and organization remain the same.

This is particularly interesting given the promise of IT work, which in many ways represents a new, positive form of work and interaction. Indeed, as I have mentioned, women working professionally or semi-professionally in IT report that they are pleased with the novelty and challenge of their jobs. And yet, in all kinds of IT work, structural social relations inform, maintain and organize their choices, participation and experiences in the workforce. Women continue to struggle to articulate the contradictions between expectations and responsibilities of domestic labour, as well as conflicting messages about their intended role in the paid labour market. And women are still under-represented in positions of power, both in terms of workplace administration and organized labour. To understand women's work in IT means that we have to be familiar with the things that remain pivotal to women's experiences.

In Canadian society, women (especially particular groups of women) have historically always worked for wages: on farms, in factories, in stores and in their own and other people's homes.[2] Women have also historically been the primary unpaid domestic labourers, responsible for housework, child and elder care and the social-emotional needs of their families, and this continues to be so.[3] Much of this unpaid work has traditionally been invisible and undercounted.[4] This division of paid and unpaid labour by gender is basic to North American society and is shaped by structural relations such as class and race/ethnicity. For example, for many working-class women, immigrant women and women of colour, domestic labour may represent paid employment as well as unpaid labour. They may work in other people's homes as well as in their own. Feminist research into women's paid and unpaid work produced the concept of the "double day" or "double shift," which refers to the domestic work that women do when they return home from paid employment. Since women perform the majority of domestic labour, this double day results in an unequal load of total daily labour between men and women. Women's greater responsibilities for unpaid housework and child care have implications for their work choices in terms of education, work hours, career path and a variety of perceived and actual employment options.

*

Brenda tells me that single motherhood was her ticket into learning how to use computers. Ten years ago, her instincts told her that no matter what job she chose, learning how to use the technology was going to be crucial. When her son was very young, she was solely responsible for his care and could not afford the daycare that would enable her to work at a paid job. To alleviate her burden a bit, she took him to the local library and while he sat happily in his play group, she sat equally happily at the library computer. Later, when her son was in grade school, she volunteered as a classroom assistant in order to gain access to the school computer. "I'd always weasel my way in and find some excuse to be in there, hacking away," she recalls with a grin. "It was a win-win situation: they got an extra hand, and I got to learn Windows."

Her home office is her base of operations. Her computer connects her to the outside world. It is always on. She tends to it like a technological Hestia stoking a hearth.[5] For Brenda, working at home is an important feature of her current job, because it enables her to have more flexibility in caring for her son. As we talk, she anxiously glances out the window for him; he is due home for lunch. Between glimpses, she notes, "This way of operating, or this type of business, I think works well for women, because work is twenty-four-seven, work now work later, it's not like it ever stops." She breaks off in mid-sentence to greet her son who has arrived and who requires guidance on what to make for lunch. They debate the merits of various sandwiches and settle on grilled cheese. He leaves to make his lunch, and I point out how impressed I am at his young independence. Brenda shrugs. "It's easy to be caught up in the superwoman I-can-do-everything syndrome: I can run my house, I can run a business out of my house, I can feed the kids, and do the laundry, and walk the dog, and, y'know, no, you can't. Because you're cheating somebody. You're cheating your boss, you're cheating your kid, you're cheating yourself. So, my son knows, when that door's closed during working hours, he'll come in and say hello, but once that door closes again, something better be on fire because I'm at work."

*

IT jobs are often non-standard in their work arrangements. Unlike the so-called standard job, which in theory involves regular shifts (such as the archetypical 9 to 5 workday), permanent full-time employment and usually some kind of employer contribution to the employee's well-being (such as a pension upon retirement), non-standard types of jobs include contract, freelance, temporary and part-time work. They may also include home-based work. These kinds of practices can both help women address the demands of their unpaid domestic labour and make labour in general omnipresent. Work can be both erratic and never-ending.

Because IT can help diminish the need for a physical office presence, it can enable women to opt for more flexible hours or to work from home. Workers can be linked to the workplace by e-mail, phone, fax, pager and a log-in to the workplace network. Wireless devices such as Blackberries or cellphone e-mail can mean that workers are always connected, wherever they are. However, the technologies themselves, while an integral part of the working arrangement, do not necessarily create optimal forms of work. Existing work practices in the IT field and employers' requirements for worker flexibility are much greater determinants of home-based work for women. Many women who would prefer to work at home are unable to do so because of the way their work is structured, while other women would rather have a full-time job in a workplace than combining multiple contingent sources of home-based income.

Working these non-standard arrangements can be a way for women to balance multiple competing responsibilities. For example, in general women often choose (or are forced to choose) to work part-time more often than men. Nearly one-third of women, compared with about 11 percent of men, work part-time. Among people who work part-time, only about 5 percent of men say that they chose part-time employment to help them balance their domestic responsibilities. In comparison, nearly 42 percent of women part-timers say that they do.[6] Working part-time is thus seen by women as a strategy for managing their domestic work. Women also turn to self-employment as another strategy to manage the demands, and are four times more likely than men to use it as a flexibility strategy.[7] Having more children also increases women's chances of self-employment and, in fact, the more children a Canadian woman has, the more likely she is to be self-employed.[8] Self-employment is a survival strategy in the IT industry, as employers have downsized workforces or simply converted full-time employees to self-employed contractors in order to cut costs.

However, despite many advantages to these types of work arrangements for women, they are not always ideal. First, employee flexibility may be quite limited in practice, especially in particular industries. Workplaces with fixed shifts, such as hospitals, stores and banks, which generally require a worker to be present during certain pre-specified hours, do not offer the same kind of time flexibility as project-oriented industries. Even in IT workplaces, which may be more relaxed, there are still deadlines, meetings and schedules. Helen, who designs the graphic interface for Web sites, tells me that although in theory she has flexibility in her work hours, it doesn't always work out in reality. Describing her day to me, she says, "I finish work about six o'clock, make dinner, spend a little time with kids, do homework, start getting them ready for bed, get them in bed, and about nine-thirty, ten o'clock, when I've finished making lunches for the next day, then I come back and work in my office till about midnight. It's very difficult to spread things out, so when a client calls you, they need the work done now, not in three weeks when you know that you've got a week in between jobs at that point." Working at home has not alleviated the demand for work to be completed according to a schedule.

Second, many part-timers find that as their hours are reduced, so too are their wages and benefits. Diane, a Web designer who is self-employed, says, "I have no benefits, and no insurance. I just try to drive veeerrry carefully." However, she is cheerfully strategic. "My partner and I are both overweight. We're both on Weight Watchers. We're hoping that we'll lose enough so that we can afford insurance." When I speak to her a year later, her business is improving, but her benefits and insurance coverage are still rudimentary and limited to basic life and critical-care insurance. She retains her cynical sense of humour about it. "If anything happens to me, it has to be death or a horrible illness," she jokes, since "minor injuries won't be good enough to get coverage." For women who are self-employed, earnings may be intermittent with periods of "feast and famine." In addition, self-employment can require long hours to generate sufficient income. For self-employed women, the flexibility to manage domestic responsibilities may evaporate in the face of increased work time.

Other women use IT to help them work at home even if their employment arrangement is a relatively standard 9-to-5 full-time, full-year job. Though non-standard work practices can lend themselves to increased flexibility for employees and the self-employed, work hours are often increased as a result because of the blurring of the boundaries between home and work. Working at home may result in the mixing of paid and unpaid labour and in the assumption that domestic work will flow smoothly into one's paid work. For example, it is commonly assumed that child care is possible while performing job-related tasks at home, and that home-based telework is a good substitute for adequate child care. But as anyone who has supervised a child full-time will know, it is very difficult to perform any but the most rudimentary of tasks while running after a rambunctious toddler. Many stay-at-home mothers of young children are thrilled to be able to find a moment between dirty diapers, feedings, cleanings, tantrums, singing Raffi songs and trying to stem the amazingly large wave of destruction that a tiny human can create to go to the bathroom or take a shower. All the technology in the world cannot ameliorate the reality of Zoodles smeared into a two-year-old's hair. As Katherine, who owns a technical services business and is mother to two children says, "Every day is a challenge between balancing family life and work. I live on a twenty-four-hour survival schedule, and I'm probably not sleeping as much as I should be. When my business started, my kids were in preschool. Even if you're freelancing, you can't freelance with kids running around demanding attention. You need to have someone giving them drinks and changing diapers."

Flexible work arrangements and the opportunity to work at home sometimes come at a price. It is often difficult for women to construct and maintain boundaries between "domestic space" and "workspace" or "domestic time" and "worktime," either because of their own work practices or the expectations of others. They have to negotiate appropriate parameters to balance their own needs as well as the requirements of their co-workers, employers, friends and family. Often women have to go through a process of establishing the legitimacy of their workspace and worktime, particularly with children. This negotiation process belies the commonly held notion that working at home is an "easy" option for women to manage paid employment along with child care. Trying to perform paid work duties at home while caring for children comes with its own set of challenges. While the homeworking women I spoke to enjoyed the freedom and comfort that this practice offered them, they also discovered that their work at home was often valued differently than paid work in a formal workplace, and that this differential valuation was often gendered.

Because of the traditional association of women with housework and child care, and because in a market economy this labour was seen as having almost no value, women's present-day home-based work can carry a subtle stigma. Despite their professional and technical status, women working at home struggle with managing competing responsibilities as well as differential (and gendered) valuation of home-based work. One of the great promises of the information revolution, the "electronic cottage," was gender-blind in its anticipation of liberation through homework. However, women's experiences of home-based IT work highlight the traditional problem of the devaluation of women's paid and unpaid work.

*

"Something about a woman working from home still sounds like 'housewife,'" snorts Brenda indignantly. "But a guy can work from home, and it means that he's just *that good* that he doesn't have to be in the office, you know what I mean? It's the double standard." She is critical of the assumptions around women's home-based paid work: that it is of less value, that it is frivolous, that it blends seamlessly into the housework that women are already doing, that it should receive lower wages because of the "fringe benefits" of free child care. "The perception problem is not on the woman's part, it's on the employer's part. Yes, on one hand, I don't have the same expenses as if I worked at an office downtown. I don't have the travel, I don't have the same wardrobe costs, I don't have to put my kid in daycare. But, on the other hand, that doesn't mean you should pay me three dollars an hour. Just because I work from home, I don't work any *less,* the quality of the materials I produce is not poorer than if I was in your office."

*

Helen also works from home. She is a veteran of the print industry, who taught herself how to use computers "sometime around the Amiga," or the mid-1980s.[9] She smiles as she recalls her 128 kilobyte Macintosh gathering dust in the basement. Currently, she is self-employed and tells me that her home-based work helps a great deal with managing the demands of her two children. But it is not a perfect arrangement, she concedes. "Working at home sometimes is good, and sometimes it's not really as good as it should be, just because everybody thinks if you're working at home, then you should be able to get this done, and can you do this for me, and y'know. Lots of life encroaches upon you." She sighs. Despite the demands, the isolation from a working community is also a problem. "You get tired of working from home and always being by yourself ... I find that when you work from home for long periods of time, that after awhile, you begin to feel like what you're doing is not as real as when you're in, quote, the office environment, where it is all business and business meetings and that kind of thing." Working at home has increased her ability to meet the demands of paid and unpaid work, but she is gripped with a vague yet powerful sense that working at home just doesn't quite count as much, somehow. Her space is always just a little too easy to invade, her work just not quite as valid. Absent from the formal workplace, she is out of sight, out of mind. Space intersects with gender to determine which work is seen as valuable. Diane adds, "It's the self-made thing that bites you in the ass. People think, why would I pay you to do a Web site when I could get Billy, age twelve, to do it in his basement?"

Listening to Helen I am reminded of Betty Friedan's clarion call, *The Feminine Mystique,* which awakened tranquilized middle-class white suburban women forty years ago. We are told that women can do it all now, thanks to technology. They are better educated and have more choices than ever before. Technologically facilitated homework has provided the solution to the problem of combining work and family. So why do some women continue to experience a discomfort that they are hesitant to articulate, not unlike what Friedan termed a "problem with no name"? Friedan wrote in 1963 that the so-called feminine mystique "emerged to glorify women's role as housewife at the very moment when the barriers to her full participation in society were lowered, at the very moment when science and education and her own ingenuity made it possible for a woman to be both wife and mother and to take an active part in the world outside the home."[10] Her discussion of the physical organization of domestic space is oddly familiar. In the home, she writes, "there are no true walls or doors; the woman in the beautiful electronic kitchen is never separated from her children. She need never feel alone for a minute, need never be by herself ... A man, of course, leaves the house for most of the day. But the feminine mystique forbids the woman this."[11] The "electronic kitchen" is a space full of the latest and greatest appliances, which, rather than saving labour, increase the demands for speed, frequency and performance. Though household technologies did not change women's responsibility for unpaid domestic labour, they often downgraded the perceived status and skill of the housewife, as the appliances were seen to do all the work. Freidan's depiction of the electronic kitchen is of a private enclosure that to the middle-class suburban housewife nevertheless remains public. Her labour is simultaneously made invisible by the wonders of household technology, and constantly on display and available to her family.[12] Like IT work in an age of piecework and just-in-time production, domestic work is reduced to a series of menial assembly-line tasks controlled by the requirements of the machinery and the demands of others. The labour expands to fill the time available despite a reduction in the time it takes to complete actual tasks.

The applicability of Friedan's thoughts on the devaluation and technological surveillance of women's domestic labour is not restricted to unpaid work in the home. In a donut store one day, I notice that a large computer screen hangs over an employee's head. As she butters bagels industriously behind the counter, the screen displays a list of incoming orders as well as a digital stopwatch that times her work down to the second. She is allocated a minute per bagel preparation. If she exceeds the time, the screen turns red to indicate the computer's displeasure, and counts the seconds of her infringement. Though the computer, not the worker, has been given the privilege of allocating work, it is the worker who bears the consequences of not measuring up to the technological standards. In theory, the technology liberates workers. In practice, for women, it can constrain, monitor and devalue their labour.

I think back further, to Virginia Woolf, writing in 1929, in a time when women were told that since they had cut their hair, learned to smoke and won the vote, they were now liberated. "How unpleasant it is to be locked out," she states, speaking of women's struggle to gain access to formerly male preserves such as educational institutions. But on the other hand, "it is worse perhaps to be locked in."[13] According to Woolf, women needed a "room of their own," a quiet, bounded space in order to do meaningful work. This space had to be free of domestic demands and had to be recognized as inviolate, a place to which a woman could escape and work in peace. For Woolf, one of the main reasons that women had not been able to achieve self-fulfillment through productive, creative labour was that the very means of doing so was denied to them. She notes that for the average woman dealing with family responsibilities, that to "have a room of her own, let alone a quiet room or sound-proof room, was out of the question."[14] While women are certainly not lacking in space that is cut off from the public sphere, finding a private space that is valued and respected is another matter. The home has the potential to be a productive space, but for women, says Woolf, it tends to be more prison than sanctuary.

The familiarity of Friedan's and Woolf's words troubles me. Why do their narratives of pre-feminist, white middle-class womanhood seem so resonant in the middle of a twenty-first-century analysis that struggles to incorporate an awareness of globalized capital and diversity? Why, in our electronic age, should the sharp tips of their pens prod my keyboard? Well, for one thing, as I type this line, I struggle to focus through a haze of interruptions. The Instant Messenger window is blinking. My e-mail beeps to tell me there is new correspondence to examine. There are renovations being performed with gas-powered machinery ten feet from my study window. In order to have a good dinner tonight I have just spent an hour preparing the ingredients to go into the "labour saving" (or perhaps time-shifting) crock-pot. The phone keeps ringing. I have the privilege of having a room of my own, but it still doesn't seem to be working all that well. Because we are able to be in constant contact with work, we are also *required* to be in constant contact with work. A Hitachi advertisement from 1997 proclaims, "Since 1960, No President Has Ever Been Out of Contact with the Rest of the World (Should You Operate Any Differently?)." It goes on to explain this with nary a shred of irony:

> There was a time when Commanders-in-Chief could disappear for days on end. They'd slip off to write a speech or fish or just make themselves generally unreachable to all three branches of government. It's been a lot of decades since presidents could do that. Heck, even you wouldn't do that. No matter where on the planet you are, you're expected to be 100% on top of it ... Whether you serve in high office or home office, you have the means to exercise the requirements of leadership.[15]

Another advertisement for a mobile hand-held computer illustrates the paradox of IT: the great thing about it is that you can work anywhere; the problem is that you are expected to work anywhere. The ad copy reads, "Whether you're writing a brief on a plane or sending email from a hotel room, it's never been easier to carry your work around."[16]

The messages that women IT workers receive still promote a gendered middle-class dream of "having it all." In the case of Friedan's housewives, having it all meant having all the latest household appliances contained in a domestic prison. In the case of the women I spoke to, having it all meant holding down a job (or several) in a tenuous economic climate while still fulfilling the domestic duties required and expected of them. It also seemed to mean incorporating information technologies into their work arrangements, so that they were constantly available to work. The message that IT workers receive is that home-based IT work is a relaxing, rewarding experience. Media images abound of happy workers nestling in their couches or reclining in their gardens with their laptops. For female workers, children and other domestic duties are added into the equation. A recent ad in *Vogue* magazine depicts a slim, chic blond woman in stiletto heels talking on a phone as her toddler dismantles her Rolodex beneath her desk. The caption reads, "A Career in Living."[17] Even if in reality such a harmonious congruence of work and life cannot happen, the myth persists that it *should.* Even if not all of the suburban white housewives of the 1960s wanted (or were able) to contentedly and smoothly manage home and family, the message was that they *ought to.* Women of the twenty-first century can bring home the bacon and fry it up in the pan. But when does the technology begin to liberate them? When they buy an Internet refrigerator that orders their groceries for them? When they can download their e-mail faster so that they can spend their time more efficiently between separating fighting kids, scrubbing the toilet and preparing dinner?

We speak often of how technology breaks down boundaries. This persisting tension between public and private space, between paid and unpaid labour, could be the unintended consequence for women. Women may not benefit from having fewer boundaries between these. If, because of the constant and still-gendered demands of domestic work, women still have to struggle to find a "room of their own," then technology does not represent a viable solution to a problem which, fundamentally, has little to do with machinery. Information technology merely provides the means for a shift in the *image* of work but not how work is *actually* done, which has changed surprisingly little. In the 1970s and 1980s in North America, women, especially groups of women who had traditionally not worked full-time, such as middle-class married women with young children, entered the job market and public workplaces in unprecedented numbers. In the first years of the twenty-first century, many of these women are shuffling back into their homes, taking their laptops with them, tethering themselves to the outside world with an umbilicus of fiber optic. But is this arrangement better? There is a risk that technologically facilitated confinement and societal organization along gender lines merely replicates the frustrations, the "strange stirring, a sense of dissatisfaction, a yearning," which Friedan described forty years ago.[18]

There is more. Because of increasing precariousness in the IT job market as a whole, because of their lower average salaries or because of the demands of their unpaid work, many women also strive to combine more than one form of paid work. Many of the women I spoke to worked freelance or contract jobs in addition to, or as a substitute for, full-time paid employment. For example, Karen is employed full-time as a Web designer, but she also puts in another twelve hours on the weekends to support her small business. Charity tells me that every weekday, she works about nine hours a day at her regular full-time job as a content co-ordinator, then comes home and does another few hours of freelance Web design work. Diane quit her full-time job because she was wearing out from the long commute, but now working at home, she puts in twenty hours of self-employed work plus ten hours chasing leads to keep her business afloat. The rest of the time, she says, she does other part-time work to pay the bills.

Since my focus was on professional, higher-income women working in certain kinds of jobs, I did not speak to women working at home in low-wage, low-status technical occupations such as data entry. But they are most definitely out there, hidden in home workplaces throughout the city, connected in electronic networks, diligently answering the phone, for example, whenever I call for a pizza. When I connect, I can hear their fingers clacking on keyboards and, sometimes, children playing in the background. The invisibility of this group of female workers is twofold: first, they are physically invisible because of their placement behind the walls of their homes; and second, they are invisible to the mainstream public gaze because of the status of their work, the interests that depend on keeping them hidden in favour of promoting the benefits of the high-tech economy, and often the colour of their skin, their family situation or their immigrant status. They have been reduced to a polite voice on the end of the phone that thanks us for calling.

However, this erasure is not enough. Now when I call for a pizza, I hear a recorded voice asking me if I want to order the same thing as last time. I have the choice to bypass a real person on the phone entirely. What happens to the keyboard-clacking women if I say yes to this too often? Lower-income technological workers, while not the focus of this book, are nevertheless implied in the relationships of the IT labour market. Much of the high-tech work is directed towards either creating or eliminating lower-income technological work. An army of lower-status IT workers may also support a few higher-status workers. Indeed, when I call for a pizza, it's usually because I've had a long day at work and don't have time to cook. The low-status work of the telephone operator facilitates my professional lifestyle. And both types of work are equally significant in our current economic condition. Both reflect major shifts in how work is done, and both reflect the differential effects of the same technologies.

In terms of professional female IT workers, it seems now that thanks to technology, these women can experience the polite claustrophobia of their middle-class mothers along with the frenzied demands of the current workplace and the modern family. On top of that they can be told that they have been liberated, that they have the freedom to mix and match jobs or even to create their own work through self-employment, and that technology should be making all of this better for them, not worse. Despite significant differences in world view and life circumstances, elements of the message that Friedan's suburban housewives received are remarkably similar to the notion of individualistic technological empowerment. You've come a long way, baby, but women's work really is never done.

The middle-class North American housewives in Friedan's work were caught in a contradiction between material privilege and personal dissatisfaction. The likewise mostly middle-class female IT workers I interviewed were also often caught in this contradiction, but their solution added a layer of complexity to the domestic work problem. Professional technical work, or combining paid jobs, meant that women in IT were working more hours to generate more income, but much of this income was used to buy back their time by purchasing someone else's labour power. For example, Melanie is a senior IT executive. Her rapid climb up the career ladder is exceptional given that when I first meet her, she is barely thirty years old. She always seems to be rushing to or from an engagement, and tells me that she cancels appointments frequently because she has double-booked. Over the time I know her, she succumbs to various forms of stress-related illnesses. I admired the sparse, tidy stylishness of her home. One day I arrive for an interview to discover the secret behind her domestic success: a young man, who has recently immigrated to Canada, is vacuuming her floor.

Another woman, Janice, a single mother who works both full-time and on contract documenting technical processes, is able to take advantage of the flexibility offered by both her employer and her own choice of mixed-work arrangements in order to care for her daughter. She is also able to take advantage of the income generated to hire a housekeeper. "I can work an hour or two so that the housekeeper can work a half day, and I can get the quality time with my daughter," she tells me. There is a certain irony in working more paid employment so that one can then pay another person to do one's housework and have time to spend on the other domestic work of child care.

One means of technological control of the unpaid work process, at least indirectly, to which women in Friedan's and Woolf's generation did not have as much access as do women in 2003, is reliable birth control. Looking over my interview records, I notice an interesting fact about the professional women I interviewed: despite their average age of thirty-five (also the average age of the IT workforce as a whole),[19] three-quarters of them are childless. I ask Donna, a Web content developer and designer, about her childless situation, as delicately as possible. Since she is sixty, I figure that she will be able to shed some light on what the younger women are up to. My question is met with a derisive snort. "My unpaid domestic demands have had no impact on my career path or job experiences. Because I didn't have children, I've always been free to pursue whatever career advantages I wanted." I probe further. But Donna is clearly uninterested in viewing the issue as systemic. She sees it as a factor of individual choice, and says sternly, "Women today need to realize there are choices which have to be made. In my opinion a working mother cannot give proper attention to either job." As part of this interview I also discover that Donna views herself as "just one of the guys." According to a recent *Globe and Mail* article, being "one of the guys" at a senior workplace level means that one is very likely to have a stay-at-home wife.[20] Since female workers like Donna are much less likely to have this extensive full-time support of a spouse, being "one of the guys" usually means making sacrifices.

*

Three hundred years ago in the West, despite the first fresh flowering of the Industrial Revolution, many people were still likely to "work from home," as the home remained the locus of economic activity. The concept of going "out" to a workplace that was differentiated from a home might have been somewhat perplexing to an eighteenth-century worker, who would likely have lived and laboured in a communal home-based enterprise, such as a farm or craft shop.[21] Those who went out to work were likely to just move to other homes. Living spaces tended to be organized around this work pattern of people residing and labouring together. In the late nineteenth century, technoscientific innovation became increasingly formalized and incorporated into norms of industrial capitalism. A well-established middle class heavily invested in notions of an essential femininity emerged. The private sphere of "home" and the public sphere of "work" became clearly differentiated for most urban people. Despite the actual presence of much paid work done at home (or in the homes of others), as well as plenty of public work in small home businesses or shops, this division grew and became more persistent in the minds of the populace.[22] In particular, the idealized image of white middle-class women as domestic angels who were not necessarily idle but who were without productive value assumed a dominant ideological status, and this ideal persisted well into the twentieth century despite the presence of large numbers of women in the labour force. Even if lower-income families could not afford the luxury of a single earner, the notion nevertheless persisted that they *should*.[23] Homes built during this period tended to emphasize this social–spatial division, and rooms were assigned for public display and private activity.

After the Second World War and the forced retirement of female workers from their industrial jobs that had supported the war effort, entire tracts of land were designed around the ideal of the "bedroom community," which would support nuclear families with male breadwinners happily commuting to their 9-to-5 jobs along spanking new asphalt highways (the metaphor of the "information highway" is no accident, as it conjures up this post-war notion of manifest technological destiny). Homes were built as little individuated boxes that lined suburban streets. In 1980, Alvin Toffler wrote optimistically of the "electronic cottage." As a result of technological innovation, Toffler predicted, work would shift back into the home and this would represent a large-scale redefinition of work. Homespace would again be reorganized around this pattern, becoming more communal, almost pre-industrial.[24] Evoking as it does a pleasant sense of the bucolic "cottage industries" (which might be more prosaically and factually termed "subcontracting"), the model of the electronic cottage suggests a warm technologically facilitated domesticity, with workers snuggled into their network security blankets.

But if women have traditionally been the primary unpaid labourers for the family, and if new forms of work replicate old ones in this regard, then *for whom* is the electronic cottage a utopian vision? Would Friedan's housewives, trapped like hamsters in a network of translucent cages, have regarded this as progress? What about women who have always done home-based work, such as pieceworkers or domestic workers? Are women working at home now in IT-based professions experiencing the electronic cottage as positive and productive? The answer seems to be yes and no. While the women that I spoke to embraced the flexibility that homework and non-standard work arrangements brought them, they were concerned with the way their work was perceived by their employers, peers and family members and with the expectations that they would continue to be responsible for the bulk of domestic labour. Despite technological innovations, the gendered relations of unpaid work seemed remarkably consistent.

Currently, Canadian women are faced with even greater unpaid work demands as a result of structural shifts in work practices. In general, these demands take two forms. First, women's traditional caring work is once again being used as a "reserve" in the face of cutbacks to health and social services. Responsibility for care of others — young children, people with disabilities or other activity limitations, elderly people, people who are ill — has slowly and insidiously been creeping back into the household as a result of government cuts, restructuring and privatization of these services. Sick people are sent home earlier from the hospital, child care is harder to find and people with special needs and disabilities may find themselves suddenly without coverage as a result of categorical redefinition. As the population ages, the need for elder care also becomes more acute, especially since women are more likely to be caregivers for ailing spouses and parents. Women are scrambling to find ways to balance these ever-increasing responsibilities. And yet provision of social services continues to erode since, it is assumed, women will be there to pick up the slack in the household.

The second type of demand, which is explicitly technological, has to do with another form of unpaid work, which Heather Menzies terms "shadow work."[25] Twenty years ago, one might have gone to the bank and done various financial transactions with the teller; perhaps deposited a paycheque or paid some bills. One might then have gone to the grocery store on the way home and picked up something for dinner. These groceries would have been checked out and packed by a cashier. Today, many of us sit down at the computer to manage our accounts and pay our bills. We visit automatic teller machines, and our paycheques are deposited electronically. When we visit the store, we might have the option of checking ourselves out and bagging our own groceries. In the course of these transactions, our unpaid work has increased. The paid work of the teller, the payroll clerk, the cashier has disappeared. In this case, new technology has not only been used to decrease the paid labour of others but also to expand the sphere of our own unpaid labour. Also in this case, the bank teller, the payroll clerk and the cashier are jobs which are very likely to be performed by women.

INDUSTRIAL AND OCCUPATIONAL SEGREGATION

When I began interviewing women in the IT industry, they would often tell me that their workplace had a lot of women in it. I thought this quite surprising at first, given that it contradicted most evidence that IT remains male-dominated. So I began asking the people I interviewed to tell me more about their workplaces. Who was doing what job? Whose posterior occupied which desk chair? Gradually, after asking many women, it became clear that the mere presence of women in IT workplaces did not signify that they held technical jobs, or senior-level jobs. Indeed, the firm of the "new economy" sounds like a very familiar place. Technical departments are overwhelmingly male. Sales and marketing departments are roughly divided equally by gender. Communications departments are primarily female. Senior management is male. The more likely an occupation within a workplace was to be female-dominated, the less likely it was that there was a good chance of promotion. In fact, I often stumped my interview subjects by asking them how high someone in their field could go. Several could not come up with a way in which they could achieve a senior management position. Many had hit the top of their career ladder in their mid-thirties, despite working in positions which they may have found enjoyably challenging. There is nothing wrong with working at a mid-level job, of course, and many people, male and female, are unprepared to make the sacrifices required for the corner office. But it seems there are no surprises in who eventually winds up in that prime office real estate. Current research supports this anecdotal evidence. Although the IT workforce is approximately one-quarter female, women are still largely restricted to non-engineering IT occupations, are not likely to be present at senior levels and continue to take home less pay than their male counterparts, even if their qualifications and credentials are similar.[26]

*

One of the most significant and clearly established themes in women's work patterns is that of their segregation into different industries and jobs. Women tend not to do the same work that men do; and when they do, they tend to be paid less, have a lower status and have their opportunities for advancement limited. The term industrial segregation refers to the distribution of workers in various industries according to gender. In other words, men and women tend to work in different industries. Women are primarily concentrated in community, business and personal services; finance, insurance and real estate; trade and public administration.[27] These female-dominated industries are those where jobs are most likely to have the greatest degree of precariousness, minimal job security, irregular hours, lower pay and fewer protection mechanisms for workers.[28]

It is sometimes said that these industries are less productive: they do not produce goods and devices, and what they do produce is hard to quantify. How do you measure the caring of a nurse, the fashion advice of a salesperson or the concern of a community worker? Definitions of productivity inherently value certain types of work more than others, and the service-based work which the majority of women do does not make the cut. And yet we know that, as I have pointed out, women do the bulk of unpaid caregiving and housework. This labour is not viewed as productive (despite the literal production of children and more metaphorical production of labour power), though people would be unable to function adequately without these domestic tasks being performed. Without a full stomach, clean clothes and a decent place to live, it's hard to be an effective worker. It seems that "productivity," while it appears to correlate with "value," is actually a very selective category of work evaluation.[29]

Occupational segregation refers to the under-representation of women in high-status and high-paid positions within an industry. Unlike industrial segregation, which is an expression of where women are found across all jobs, occupational segregation is an expression of where women are found within one particular job group. For example, where industrial segregation describes women's under-representation in manufacturing as an industry relative to all other industries, occupational segregation describes women's access to positions of power within manufacturing alone — which jobs they hold within manufacturing workplaces, how much they are compensated, how much status they hold in each job and so forth. As with industrial segregation, occupational segregation results in women holding primarily low-status, low-paid jobs. Not only are women cloistered into a small number of jobs but they also dominate these occupations in terms of overall numbers. The more likely a woman worker is to have a particular job, the more likely it is that that job category will be female-typed and contain a higher proportion of women.

Segregation of women's work is not confined to gender, though this apparent binary is a convenient, if too-general, distinction. Rather, it is essential to look at differences among women. While women as a group are found in certain industries and occupations, there are also significant variations in women's experiences. For example, as Heather Dryburgh notes when speaking of immigrant women in the IT labour force:

> As *women* in IT, they are part of a minority in a male-dominated sector; as *immigrants* they are vulnerable to experiences of delayed employment, lower pay, and undervalued credentials and experience; ... for those who in part comprise the 75% of immigrant women in the 1990s who are *visible minorities* there are additional disadvantages to face in the labour market, including lower earnings and higher rates of unemployment.[30]

Currently, immigrant women comprise 8 percent of the IT labour force as a whole, but comprise one-half to one-third of the lower-paid, lower-status positions in IT manufacturing industries and the data processing field.[31]

Age is another important factor in women's experiences in IT. Older women often find it more difficult to secure employment in and feel a part of a culture that privileges youth and hipness. As Susan, who is fifty-eight, tells me, "Since I'm older, I often feel awkward at technical gatherings, because I'm the only woman my age. It feels like this field is more for younger women." Maria, fifty-three, says that she was surprised to be considered for her position as a content and information design manager at all. "For me to go into a company where the CEO is twenty-nine, and he's one of the oldest people on the payroll, well, y'know." She laughs. "But although I felt that my age would be an asset to his company, it was clearly a surprise to his clients. I went to a client meeting, and the client was clearly quite shocked, and he said 'I didn't expect someone your age.' They had developed this company personality based on the youth of their employees." While the average age of an IT worker is thirty-six, many professions such as Web design draw from a much younger group. The average age of a Web designer is thirty-two.[32]

Finally, geography is another significant disparity among women: the experience of working in Toronto, a large urban centre, is qualitatively different from smaller Canadian cities, industry towns and rural areas. Regardless of a woman's technical aptitude, if she is located in an area where the only jobs are at the local factory or fish plant, then she will have difficulty securing high-tech employment. Despite the possibility of virtual work, physical geography remains a significant constraint. Two-thirds of IT workers are concentrated in five major urban areas, including Toronto.[33]

This industrial and occupational segregation may be responsible for the somewhat non-linear career paths of many women in IT. It would seem logical that a person with technical aptitude and interest would simply enroll in technical education, then upon graduation locate a technical job. Yet this was, in fact, the least common path into IT work for the women I spoke to. A more common trend was for women to find themselves in IT work through unconventional channels. Many had stumbled across it quite by accident. They had been involved in other, often female-typed fields and had been impressed by the opportunities that the technology offered. Some women started quite young, some became interested in technology as a result of their education, but most did not follow a linear career path from formal post-secondary instruction to IT work. Interestingly, these experiences did not necessarily correlate with the current job these women held. For example, a Web designer who worked for a large technology company had originally dropped out of theatre courses in university. A senior level network engineer had barely passed high school and had dropped out of English literature courses in first year university (because, as she noted, "Computers pay better than Chaucer"). Conversely, a woman who had had formal science and technical instruction since childhood (and had completed a master's degree in the field) was working in a government health-care agency, while another who held an applied science degree was employed by a marketing firm. These types of "hybrid" jobs, which combined various tasks and occupational characteristics, seemed to be characteristic of women's work in IT. In part, this is perhaps due to the segregation traditionally experienced in the labour force.

Yet among the women I interviewed, although the content of their work was fairly technical, many of the jobs themselves were largely offshoots of traditionally female job areas: writing, editing, communications, public relations, marketing, office administration, customer service, content management, librarianship. Although they might have been employed in a fairly technical capacity, they still experienced segregation on the job. These women were technically competent, but only a few were programmers, engineers, hackers or software designers; only a few of them spent all day with their hands inside a computer case or connecting the office network to the coffee maker. There are some marvelous exceptions, of course. For example, there is Laurie, a whiz with machine language, who dabbled in robotics and who is skilled in creating the control systems for automated industrial production. There is Lilith, who spent most of her teenage years running a bulletin board, programming and hacking, who had worked for NASA and who was presently considering how to hook her television up to her wireless home network so that she could record digital cable programs remotely. There is Nadine, an older woman who had specialized in repairing mainframes in the 1970s and 1980s. And there is Erin, a former zoologist who entertained me with her descriptions of "percussive maintenance" on computers: "The early Apple IIs buzzed and vibrated a lot, and the internal card would wiggle its way out. So you had to take it apart, and then you'd pick it up, from a height of about six inches, and you'd go BAM! and drop it on the table. It would reseat the card and then it would work." But in most cases, when I asked, "Who makes technical decisions?" or "How many women are in the tech department?" the answers clearly demonstrated a persistent occupational segregation. For example, Carmen, an information architect, a profession associated with librarianship, tells me, "There are eight information architects, and one is male. All the rest are female. There are about ten tech leads, and one is female."

Women are making some inroads into male-dominated, high-status positions, but in part this is because much of this work no longer resembles the more traditional male-dominated professional work. In the case of IT, this appears to be true: as certain types of IT positions becomes less associated with "hard" technology and academic computer science, more women appear to be performing them. While the technology has not necessarily changed, how the work is valued has certainly shifted. The numbers are not enough. Just adding women into technical workplaces does not change the occupational dynamics that have persisted for decades. "Add women and stir" is not a recipe for dramatic social transformation.

This shift in occupational and worker characteristics, this change in who is doing the work and how it is valued, is often referred to as the feminization of employment. In general terms, feminization of employment means two things: that women are in the labour force in greater numbers, and that many jobs are increasingly beginning to resemble "women's jobs" — service-oriented, often precarious, part-time, temporary and low-paid.[34] In the case of IT work, feminization has taken on a few different forms. First, while women are moving into IT jobs, they are doing so in part because many IT jobs no longer resemble the types of male-dominated "hardcore techie" jobs of ten years ago, but are often rather hybrid types of work that combine elements of more familiar jobs that women have traditionally done. Second, while women are in IT work, they do not tend to be found in the IT professions that carry the most value, seniority or compensation. Indeed, the polarization of jobs that tends to accompany the process of feminization has also occurred in the IT industry. There are a few well-paid, secure professional positions, but there are a lot more contingent, sporadic, lower-waged, benefit-free jobs and women, particularly younger women, immigrant women and women of colour are more likely to be found here. Finally, the movement of women and many groups of immigrants into the IT labour force has been accompanied by the devaluation of many of the jobs they do.

*

When I ask Brenda how she feels about working as a woman in IT, given the traditional association of IT with white males, she is dismissive of the idea that only certain types of people should work in technology-related professions, and indicates vehemently that logic is only part of the IT equation. "I've met enough people in this business who are mostly computer people. They write code, or applications, whatever. And I think to myself, you really should stay in your basement, because you have no social skills to speak of. You shouldn't come out here unless you're prepared to deal with us illogical humans. Computers are so literal. People need a bit more *work.* We need interpretation. Our feelings *matter.* With a monitor, you just yell at it." To illustrate this point, she demonstrates, hands flailing, a raucous ululation of *ayayayaya.* "I compare those people to Vulcans ... To them that's a compliment, and in a way it is. They're highly logical human beings. A leads to B leads to C leads to D. Well, humans are not like that. That's why I say this area offers so much variety. Are you a highly logical, mathematical, left-brain person? Here, learn a bunch of code, and write your own stuff or whatever. Have no interest in that, are you more a salesperson, a more babykissin', handshakin' kind of person? Fine, then go sell what he just wrote. There's every type of experience that anybody could want."

Carmen agrees, and suggests that even though men and women may be interested in different things, it's the compensation and valuation that matters. "Maybe we are just attracted to different things and have different skills, and maybe they just need to be *valued* equally. That's what we need to fight for, rather than just saying, I can do anything you can do. I can do anything, but I don't *want* to do anything, I want to do what I'm interested in."

*

Despite many shifts in work and a move towards employment feminization, "hard" technical industries and occupations remain male-dominated. And yet this cannot be reduced entirely to gender. In the case of IT, there is another dimension to this arrangement. There appears to be a fascinating contradiction in IT workplaces, which demonstrates the importance of looking carefully through a gendered lens at how work is valued. That contradiction is this: while technical skill is clearly correlated with masculinity, and tends to be rewarded over the so-called soft skills that are associated with femininity, some groups of technical males are not as well rewarded as others; in fact, the more likely some jobs in IT are *purely* technical, the less likely it is that they carry job status. As I write this, I think of the technical department within the building where I work on campus. Like many stereotypical IT departments, it's windowless and in the basement, isolated from the rest of the building. There are two women who stand out like periscopes among a sea of males. In general, while faculty may depend on these people for computer support, they don't view them as intellectual or professional equals. They're sort of like the IT equivalent of plumbers or caretakers, someone you call when something goes wrong. Despite these men doing primarily mental, not manual labour, this particular strata of technical males appear poised to assume a new kind of working-class masculinity. Like the unionized trades, these jobs are generally male-dominated, reasonably well compensated and require a fair bit of experiential knowledge and technical skill, but they are not necessarily valued as professional.

Nonetheless, the technical department can hold non-technical workers in thrall. A university colleague tells me of a pre-eminent female scholar whose intellectual credentials are beyond reproach and whose publication record is as long as my arm, but who was nearly reduced to tears when the man who came to fix her computer hinted disparagingly at her technical ignorance. I engaged in battles with several men from the technical department over my choice of desktop operating system. Frustrated with the peculiarities of Windows, I investigated some solutions and eventually asked that the Linux operating system be added to my computer. The tech department was clearly interested in maintaining epistemological and software superiority and treated my request with the same dismissive hostility, as if I had asked one of them to squish a live kitten into my CD-ROM drive. The fact that I was university staff, did research on technology and held a PhD meant nothing; I was young and female and not part of their community of knowers. They were the technical experts and they would dictate the rules. Finally, one of the gentler men relented and helped me make the changes I wanted. The atmosphere of illicit secrecy that surrounded our insertion of a contraband video card gave the O/S switch the aroma of a dirty affair.

Thus, although gender is a very significant element in industrial and occupational segregation, this phenomenon cannot be reduced to gender alone. In the next chapter, I take a more in-depth look at the creation of technical communities and how these communities are structured by gender, age and perceived skills.

PAY DISPARITY

Women, as a group, are paid less than men. In the two decades following the 1970s influx of women into the labour force, the pay gap had been shrinking as women gradually moved into higher-paying professions and joined unions, and as men's well-paid jobs declined. But currently, the pay gap between Canadian men and women appears to be growing again. Women with university degrees, who traditionally had the best odds of commanding higher salaries, are now earning 69.8 percent of what men with university degrees earn. Shockingly, this has *declined* from a previous high of 75.9 percent in 1995.[35] Part of this pay gap can be explained in two ways. As shown in the previous section, women and men tend to work in different industries and occupations, which are differently valued and remunerated (and this valuation is based on criteria that are themselves shaped by assumptions about the value of work). The sectors and occupations that are female-dominated tend to be lower-paid than sectors and occupations that are male-dominated. In addition, women are often paid less than men for performing exactly the same work. The combination of these two factors results in a net effect of women, on average, earning significantly less than men. When other factors are included in the equation, such as domestic labour load and possible reduced income as a result of alteration in employment practices to accommodate childrearing, it becomes apparent that women are generally at a significant economic disadvantage relative to men.[36]

Structural and systemic relations, as well as individual factors, fundamentally shape women's diverse experiences in the salaried labour force. These factors are not accidental or "added on"; rather, they are an intrinsic part of the organization of work in our society. For example, many employers in Canada have historically depended on the cheaper labour of women and immigrant workers. Immigrant workers who are women would thus be expected to make the lowest wages, and indeed there is a disparity in women's earnings by immigration status, including "visible minority" status.[37] There is also a disparity in women's earnings that depends on the timing of their childbearing, with women who have children early earning 6 percent less on average than women who delay childbearing (or eliminate it altogether).[38] Interestingly, according to one researcher, "*roughly one-half to three-quarters of the gender wage gap cannot be explained.*"[39] In other words, once one has accounted for skill, education, experience and other measurable factors, there remains a persistent wage gap that does not appear to depend on these things.

The data on wages from the IT industry paint a familiar picture.[40] In 2000, the average wage gap between men and women working in IT-related occupations was about 80 percent,[41] but wage disparities vary widely, depending on the job. For example, female computer and software engineers can expect to make between 75 and 79 percent of their male counterparts' annual earnings. The few women who might supervise these engineers would, in fact, see the wage gap increase to an astonishing 62 percent as they moved up the ladder into senior management. However, women doing some jobs such as systems analysts and computer programmers could make up to 90 percent of what men earned for the same job. Yet, as we have seen, since women tend not to be found as often in these positions, any advantage that they might have as a group evaporates in the face of their work segregation. Sarah, a Web designer in her twenties, is excited about her new career. She tells me authoritatively, "A good job in IT is so empowering because you make so much more money. Money is power, in this society." Unfortunately for her, her chances are not good. A third of women in IT are Web designers. Web designers work some of the longest hours of any workers in the IT industry, but they make some of the lowest wages and have the highest unemployment rate.[42]An economic disadvantage represents a social disadvantage as well. A good job in IT can, in theory, be one route to economic self-sufficiency for women. However, it can also replicate larger problems that women experience in the labour force.

The Role of Technology

It is undeniable that work discourses and practices have changed in North America historically and that this change has happened (in part) in relation to the development of various forms of technology, particularly industrial/manufacturing and information technology. While this development has been lauded by some, particularly business leaders, it has brought criticism from others who see technology as responsible for negative effects in the workplace.[43] Technologically facilitated work, says Ursula Franklin, ironically despite its mantra of connectedness, has in fact profoundly disconnected us from the labours we perform.[44] By separating the process of work from the product and breaking the process itself into tiny, manageable steps, technology has created a "culture of compliance."[45] Interestingly, compliance is also seen as a virtue in women and marginalized groups: sit down, be quiet, accept your place, don't make a fuss, don't be bitchy or pushy, be grateful for what you have.

However, although it is tempting to see technology as the cause of change in the workplace, this conclusion is too simplistic. It does not adequately address the complex web of relations that is inherent to both technology and work. Technology as a set of objects does not inherently determine its own use. Rather, use is incorporated into human relationships and structures in particular ways based on the fulfillment of certain needs. Often, as a result of this process, technological objects will be employed in struggles over their use, or their eventual use will end up being contrary to their intended use. We all know the telephone as a familiar, perhaps feminized, household appliance. Yet, in its early days, it was viewed as solely a tool for enlightened businessmen; women might be able to use it but it would likely be too complex for regular household use. Technological objects and practices are hence inseparable from their social context and shape one another in an ongoing, dynamic process. Given that both work and technology are crucial to defining, constructing, maintaining and reproducing structural social relations, any analysis of workplace change through technology must be viewed through these social systems and their ideological as well as material causes and effects.

It is important not to make sweeping generalizations about women's work in and with technology. Like gender, technologies have histories and diversity. Many earlier analyses of the role that technology plays in shaping work did not emphasize the individual and social variation among workers. Some visions of the future are gender-blind, while others homogenize women (and technology).[46] Analyzing women's work in IT is challenging precisely because it cuts across several categories and resists easy classification. The experiences of women with IT are contradictory, full of surprises and creative tensions, provocatively suggesting new areas for work and research.

For the women profiled in this book, IT work is a lens that focuses their identities and experiences. Technology *is lived in and through material social relations:* of work and organization, interpersonal interaction and structural dynamics of institutions. For these women, ideologies around technology are not abstract, but a product of lived experience. Similar experiences are not interpreted in the same way by all women, although like circumstances often result in shared perspectives. Though there are discernibly gendered trends in women's work in IT, it is essential to produce a grounded, context-based analysis of the ideologies, discourses, objects and material practices of this work, as well as how social locations like gender are described, maintained and prescribed in the process. Technological change does not happen in isolation. It is mediated through structural power relations as well as deliberate choices within the workplace.[47] In the next chapter, I will examine how these structural power relations are applied to the recognition and evaluation of skills, particularly technical skills.

CHAPTER 2

The Struggle for Skills

LINA IS PISSED OFF. As she describes her experiences, her voice becomes an angry staccato, and she gestures wildly to give her words emphasis. She is surrounded by the accoutrements of a bored woman: little craft projects requiring dexterity and meticulous attention, a TV still warm from an afternoon soap opera, an impeccably tidy and carefully well-decorated apartment. She grimaces, "I'm an expert in cooking, cross-stitch, painting ... I have so much energy and time." Lina wants desperately to work, but she can't. Lina is in Canada because she fell in love, and at this point, her attachment to her Canadian husband is a tenuous thread of connection to this country. Born and raised in Mexico, Lina received a BA in Communications and an MA in IT Management from a prestigious university she calls "the Mexican equivalent of Harvard." During her graduate studies, she developed virtual learning communities for IT companies and created and implemented the systems to run them. In Mexico, she says, she was "aggressively recruited" by top IT firms and worked as a telecommunications consultant for a large Mexican IT company. Confident that with her qualifications she could find work in Canada, she agreed to move.

When Lina arrived in Canada, she had to wait several months just to get permission to work. Undaunted, she applied for jobs everywhere as soon as she was able. She also enrolled in a Toronto university in order to obtain additional technical credentials, and was underwhelmed by the quality of Canadian education, which she felt was inferior to that available at Latin American universities. When she finished her courses, Lina was at the top of her class. After many months of searching, she became disillusioned as employer after employer turned her down for lack of "Canadian experience." "I can get all the hundred percents at school I want," she says, "but it's not going to get me a job." She began noticing that patterns emerged in employer treatment. Other Mexicans she knew stopped telling employers of their country of origin, and only then were they successful at landing employment. Employers were able to circumvent employment equity provisions by invoking "Canadian experience." She managed to land one job interview for a good job in her field, and even though she agreed to work shifts for it, the company rescinded the position after the interview and offered her a receptionist job instead.

Now she is both confused and angry. Before she came to Canada she was told that there would be plenty of employment opportunities for her here. She passed the "points" test, which evaluates the suitability of skilled immigrants, with flying colours.[1] She is puzzled about Canadian companies' lack of interest in a job candidate who is well qualified, seen as valuable in her home country, is personable, eager to learn and pleasant and who speaks two languages fluently. "Why would you not take advantage of people who speak Spanish and have a good education?" she asks. This rhetorical question is set adrift in my direction, but it is clearly aimed beyond me: at a labour market whose rules are inconsistent, at a government who misrepresents its nation abroad, at immigrant recruitment practices that don't deliver on promises, at employers who toss out her résumé because of her name and birthplace, at a society which is insular and unaware of its position in a global village. "If you don't need people," she continues, "why ask people to come? They are lying to us! Now you're here as an immigrant, maybe you sold everything, said goodbye to your family, brought enough money for six months, and you wind up with nothing!" She is starting to wonder if, in fact, it is possible for an immigrant, visible minority woman to do more than work menial jobs, and notes sardonically, "What you want are high-skilled servants. Do they think we should all be cleaning or serving in a restaurant?"

She gazes out the window of her apartment, into the wavy haze of the city summer air. "I feel so empty without work, so desperate. I don't want to wake up in the morning. I don't see a future here."

*

As I ride the subway or walk through the streets of Toronto, I am often confronted by a series of advertisements that feature young, attractive, multi-racial, multi-national people who have improved their lives by improving their IT skills. These ads tell me that these people have forsaken thankless jobs as hairdressers or store clerks in order to pursue a challenging, exciting, well-paying career in IT. All that lies between drudgery and opportunity is a certificate from a private IT college to indicate that the prerequisite skills have been acquired. It is almost as if these colleges are performing missionary work amongst the heathen, who once were lost, but thanks to MCSE or CCNA or a host of other alphabet-soup certifications, are now found.[2]

Skills is a buzzword in the IT field, and the relations of skills in the IT industry seem quite straightforward. IT workers need to have skills, to upgrade and promote them. They need to have both "hard" technical and, increasingly, "soft" interpersonal skills. They assume that they will be adequately compensated for having invested in their skill development. Government agencies produce earnest rhetoric about the "knowledge skills" deficit of the Canadian workforce, provide incentives for skills acquisition and initiate programs aimed at matching skilled workers to employers. Ironically, it appears that the skilled workers whose abilities Canada needs most are not mental but manual labourers: tradespeople like carpenters, electricians and plumbers. In material generated by the government and mainstream media, there appears to be a clear relationship between skills and IT work quality and compensation. The more skills you have, the better your work experience will be. A skilled workforce is valuable, and thus we must strive to improve our skills.

It would be foolish to argue that skills are useless. A senior programmer is undoubtedly better off in the IT field than someone whose technical skills are limited to making Kraft Dinner in the microwave. Many employers are in genuine need of workers who know how to perform important and difficult tasks. However, we see throughout this book that gender stratification in the IT workforce is a common theme. It is also clear that Canada hosts a large army of underemployed immigrant workers. There are computer programmers from Lebanon driving cabs, biochemists from Somalia shining shoes in the airport, lawyers from Nigeria doing data entry, doctors from

China working graveyard shifts in convenience stores. Job status and value do not depend entirely on the skills that each individual acquires. Are women simply less skilled than men, immigrants less skilled than Canadian nationals, "outsourced" or "downsized" workers less skilled than those who have managed to cling by their fingernails to their jobs? The evidence tells us that the answer is more complicated than a simple yes.

In this chapter I explore the relations of skills and how they play out along particular lines in the IT field. Despite rhetoric of economic and personal empowerment through individual skills acquisition, there is only a tenuous link between skills and job compensation and status. What counts as skill, and what is valued as skill, is mediated by social location, that is, the complex interplay between individual and social characteristics such as gender, race, class and so forth. Even if one has skills, they may not be viewed as the right kind of skills, or the same skills may be disparately valued. Additionally, the path to acquisition of the desired skills may vary depending on one's access to institutions. Workforce stratification, contrary to conventional wisdom, does not depend entirely on "objective" skills, which have empirical value. Rather, like commodities or currencies that have no inherent value until they are given it within a market system, skills are evaluated through a contextual web of systemic dynamics of power and privilege. In other words, there is no intrinsic value to cowrie shells, stones with holes in them, slips of paper or pieces of metal with Queen Elizabeth's face on them, until we all agree that it should be so. Likewise, there is little inherent value in many skills until a social consensus, and the labour market, assigns it to them. What is considered valuable can change. Certain IT skills acquired sudden value in the 1990s because so few people possessed them. Dorky hobbies performed by dweebs suddenly became vital activities performed by tech celebrities.

But what investment could the IT industry have in an unequal valuation of skills? Surely this industry provides us all with a field of democratic possibilities, which are determined only by our hard work and perseverance in acquiring skills. This "pull yourself up by your bootstraps" culture of the North American IT field is not accidental, but is derived from a social, political and economic agenda that has a vested interest in the promotion of individual self-empowerment. There is nothing wrong with hard work and perseverance, of course. They are wonderful, useful qualities. But reducing success entirely to individual factors such as skill not only obscures relations that occur on a larger scale (such as workplace hierarchies of advancement) but also prevents the consideration of alternatives. If we make less money, fail to acquire secure full-time employment or are discouraged from promotion, then it must be entirely our own fault: we have failed to tap into our "person power." Reducing complex systems of power and privilege to individual factors hides the operation of social structures and eliminates even the tools to contemplate a workplace in which this is not so. If we possess no language or conceptual strategies to think about these systems, if we feel that we are the only one who is unable to succeed (and that this is entirely our own fault), if we reduce ourselves to "human capital," our skills to quantifiable commodities and our education to consumption, then we are unable or unlikely to challenge relations of inequality that operate in our lives and the lives of others. Put another way, the mantra of individual achievement privileges skills acquisition and "personal empowerment" as the primary means of advancement in the field. As a result, social location, a key factor in women's labour force experience, is discursively erased and deliberately obscured. This places the onus on the worker to understand her own success or failure in individual terms, which have no connection to structural relations, hide the systems of power operative in the workplace and prevent activism and organization on the basis of shared concerns or identities.

A SKILLS SHORTAGE?

In recent years we have been presented with the notion that there is a skills shortage among Canadians and that this is impeding our full technological development. For example, a government document produced by Human Resources Development Canada lists as one of its goals the increase in "knowledge infrastructure" and argues that "more of Canada's private sector needs to aggressively develop its capacity to create new ideas and bring them to market."[3] This document suggests that a drop in national productivity in a "knowledge economy" is the fault of a lack of "innovation capacity." From the private sector, a report on the status of Ontario's technology field, commissioned by the Information Technology Association of Canada and prepared by Aon Consulting, states that there is a definite shortage of "IT-qualified professionals" in Ontario,[4] and that "58.9% of companies experienced a 'measurable loss' as a result of shortage of IT skills."[5] It would appear that the acquisition of technological skills is a collective responsibility and that our economic viability depends on it. Yet a recent Statistics Canada report that examines the labour market of the 1990s expresses some uncertainty about the role of technology skills in employment practices. It noted that although there was a strong demand for skilled technology workers, compensation for these workers was not evident. Earnings for young people declined in the 1990s, even though they were most likely to possess desired technical skills. Thus, the report concludes, "there are important phenomena that cannot be explained and are at times inconsistent with the belief that technology is the sole, or perhaps even primary, driving force."[6] The recent downturns in the technology sector, including the spectacular implosions of various high-tech and telecommunications companies as well as massive job losses, cannot solely be blamed on a lack of technological skills. Indeed, at times it would appear that over-valuation of technical skills might lead to inappropriate business decisions.

There appears to be an inconsistency among the apparent skills shortage, massive layoffs, cuts in the IT industry and declining salaries. Indeed, as a 2000 Canadian government report indicates, "There was no evidence of a generalized shortage of technical skills immediately threatening the ability of Canadian firms to compete in global markets," but more significantly, that "only a small minority of firms reported raising salaries to attract the people they need … Based on wage data, there seems to be little basis to conclude that the ICT sector is starved for talent."[7] Moreover, concluded the report, "of far greater concern than a lack of skills is a shortage of opportunities in Canada."[8] A 2001 report by the Organization for Economic Cooperation and Development reaches the same conclusion and calls the claims of severe skills shortage "unsubstantiated."[9] What, then, are we to make of this disparity between policy recommendations, ideologies of skills acquisition and the reality of the IT labour market?

Human Capital Theory

The IT sector has been deeply invested in the notion of meritocracy, or the idea that people succeed or fail based entirely on their own individual capacities. In the culture of information technology, there is a strong message that skill, and skill alone, determines the performance of individuals. Often, this value given to skill possession and use has a macho element of posturing and competitiveness. Like kung-fu B-movies where acolytes of various styles battle for supremacy with plenty of teeth baring, trash talking and skirmishes in which to demonstrate their abilities, IT culture is rife with challenges to technological manhood. Indeed, IT culture is often represented as a sort of techno-Darwinist environment where skill is the claws and teeth that provide the evolutionary advantage. In practice, there are also strong currents of collaboration and community, knowledge sharing and mentoring. Nevetheless, the pervasive ideology of skill supremacy and individual competition proliferates. Jane Margolis and Allan Fisher describe the culture of computer science at Carnegie Mellon University where "distinctions are argued all the time: 'Real' programmers use C; command-line interfaces are better than GUIs; 'Macs are for wimps.' A professor teaching a programming course in C++ derides the 'stupid languages', such as Java ... He calls Java a 'programming language for bozos.'"[10] Numerous theorists of women's technical education, including Margolis and Fisher, point out that women tend to be implicitly or explicitly excluded from the community of technological experts. This affects both women's own perceptions of their performance and competence and the larger culture of IT education and work. Though not all men are included (indeed, it is clear that the culture is based on active *exclusion* of non-technical males and on *hierarchies* among technical males), skill supremacy is viewed as inherently masculine and imbued with macho values.

The premium placed on skill supremacy and so-called human capital in the IT field was initially what distinguished it from other fields. Other, more established fields of work, while they also respect skills, tend to recognize that elements such as networking, personal charisma and strategy also ensure success. Perhaps the emphasis on skills acquisition in IT culture results from the experiences of workers who felt that as "nerds" they did not have access to these social resources.

Perhaps also the concept of human capital is attractive in a neo-liberal social climate that privileges the logic of free market capitalism and discounts the workings of systemic social inequalities. In any case, this mentality has become pervasive and now informs larger policy agendas outside the IT field.

As a result, we must critically investigate the claim that having skills leads automatically to "progress." We might ask, What skills are actually needed and who is able to possess these skills or have them recognized? How are these skills implemented once they are acquired? And above all, does the promise of reward for skills hold up? While I hesitate to incorporate a theoretical discussion into this book, which is meant to be primarily about women's experiences in the IT labour market, I think it is important to have this conceptual chat, because the ideological premises of what is called *human capital theory* reflect much of the mindset of corporate IT culture. In part, this reflects a perception of IT workers as highly individuated free agents in a neutral open market, rather than embedded in systems of power and privilege.[11] More broadly, this perspective of intellectual entrepreneurial individualism is also guiding the public policy decisions of government and business organizations in Canada, as well as the general public opinion and the current political climate in Ontario. The tenacity of human capital theory as a model that explains worker experiences is particularly problematic at a time when there is a "knowledge based economy"; though more of us are more highly educated than ever before, more of us are also working at jobs of poorer quality, or not working at all. Despite rising levels of education among the general population, there is increasing precariousness in the labour market as a whole: good, secure jobs are disappearing and being replaced by temporary, contract or part-time jobs, often in the service industries and usually without benefits or employee security. It seems that our job quality is decreasing as our certifications pile up around us.

This theory is particularly insidious because it blames individual workers, not larger economic or social factors, for this apparent employment failure. In my interviews with women IT workers, I was interested in the way that they often explained their failure or success in terms of their own human capital, but few of their experiences seemed to match the actual amount of human capital they possessed. If they lost jobs, it was not because of their lack of skill but more frequently because of corporate downsizing; if they were hired at jobs it was not necessarily their qualifications that got them in the door. Nevertheless, they struggled to put their often contradictory experiences in terms of the common logic of the corporate IT culture and field. They worked hard to build up the educational credentials they thought they needed. Unfortunately for them, this did not always translate into promotions or into landing good, secure jobs.

To return to my original discussion, human capital theory is one explanation offered in traditional economic models for the salary and job status disparity between people in the labour market.[12] It is a theory steeped in optimistic and atomistic economic models of the capitalist free market, though ironically the term first appeared in 1930, surely a time when people were less inclined to feel that their economic situation had much to do with their individual characteristics. In these idealized theories, workers organize themselves neatly into productive units like cogs in a well-oiled Swiss clock. Each person is akin to a mini-venture capitalist, an intellectual investor with equivalent resources at her disposal and the freedom to play the educational stock market where she chooses. "Human capital" refers to the amount of individual "investment" each individual makes in herself to make herself more attractive to the labour market. Capital is invested in the form of education, training and development of skills, and individuals then reap the dividends of their investment in the form of higher status and higher pay. In this meritocratic model, jobs with higher status and pay are a direct reward for personal investment. Women, because of their unpaid labour demands, particularly childrearing, as well as their personal choices about education and training, are thought to have lower levels of human capital, and this is the reason for their under-representation in well-paid, high status occupational categories and jobs. Moreover, the theory holds, women deliberately choose jobs that require less human capital and lower productivity because of their anticipated domestic demands.[13] Skills, then, are things that one rationally chooses to acquire (or consume, we might say) as a commodity, based on one's consideration of one's own needs and on the needs of the labour market.

Human capital theory is appealingly simple, fits well into capitalist models of free-market investment and return, and seems to offer us some clue to women's lower job status and salaries in the IT industry. It tells us that women make less and are valued less because they have fewer skills. The theory also tells us that if women seek to earn more, then all they have to do is acquire the same skills and work as hard as their male counterparts, and full equality will naturally follow. However, as with all theories that are just a little bit too neat and tidy, there are some problems. The first significant problem is the complete erasure of other variables and structural determinants than gender. As I have stressed throughout, several factors combine to shape women's diverse experiences in the workplace. Particular groups of women, and particular occupations, are more likely to be lower paid. Historically, patterns of work organization have depended upon dividing the workforce according to social location. These divisions are not accidental but are derived from deliberate choices by individual employers, state policies and a host of intersecting systemic power relations. In earlier periods, employers were quite explicit about hiring particular people because they knew they could pay them less.[14] At the present time, there are more social taboos against openly stating preferences for certain groups of workers, but the practices continue.[15] This results in not only the production of groups of workers who are marked by social signifiers but also in the reproduction of these relations.[16] This production and reproduction operate overtly and covertly both at the individual "everyday" level as well as at the institutional level, which includes the state's participation.[17]

Like Lina, many immigrants to Canada during the last ten years, despite being on average more skilled than Canadian-born workers,[18] have had the unhappy experience of having their skills devalued. They arrive with various types of education, training and knowledge only to discover that they are unable to use them and that their skills may not be accredited. This is particularly interesting in light of the current trend to "outsource" labour to other countries (for example, India) because of their skilled workforce.[19] Indeed, the contradiction between the practice of contracting out labour to other countries and the practice of denying jobs to qualified immigrants in Canada is one of the most significant demonstrations of the socially constructed hierarchy of skills. What makes a workforce in a country like India appealing are its skills and education, but more importantly, the ability of the employers to pay the workers less than North American workers. Move the same workers across the globe to Canada, and suddenly their skills do not translate. Even if they are able to find a job that demands their skills, immigrant workers are paid less than Canadian-born workers and immigrant women, especially immigrant women who are also visible minorities, are paid the least of all.[20] The obstacles to compensation and job status are clearly structural in their close ties to Canada's immigration policies. Currently, the Canadian government is exploring ways to facilitate the entry and employment of immigrants who possess the desired skills. Interestingly, the IT field may represent one area in which immigrant workers may make good inroads, since formal credentials are typically less valued in IT than in other fields and on-the-job experience is considered more important.[21] However, given the experiences of people like Lina, it remains to be seen whether emphasis on formal skills acquisition and the elusive "Canadian experience" will diminish this window of opportunity.

A second problem with human capital theory is its emphasis on individual characteristics. It assumes that the labour market operates in direct correlation to the innate or gained personal attributes of individuals. This emphasis on individual characteristics is undermined somewhat by a concomitant focus on the "objective" laws of supply and demand, which often accompany human capital explanations. But more significantly, it places the blame on individuals for not achieving high status. It faults women for their own under-representation in positions of value and avoids structural relations of power entirely. Yet extrinsic, not intrinsic factors, are more likely responsible for a perceived lack of appropriate skill.[22] In other words, how people view my skills may have more to do with them than with me. The women I interviewed often stressed that employers' and co-workers' perceptions of their skills did not match what they actually had. Many had to work twice as hard to be considered half as good.

Despite a focus on individual characteristics, non-quantifiable individual factors such as sex-role socialization and perceptions of social expectations are not included in the human capital analysis.[23] Perhaps as a result of poor role modelling and reinforcement of socially sanctioned gender expectations, women themselves consistently underestimate their own skills. One of my interviewees, Charmaine, tells me about her own experience with devaluing her own abilities. She wanted to join a women's technology organization, but although the avowed mandate of the organization was to help women feel comfortable with technology, Charmaine felt as if she needed to "qualify" to be a member. Even though she "lived on-line," she decided that she lacked the competence even to be part of an organization for technological activism. "I thought, oh, I don't know enough to join. I don't know why I would think that, but I guess I felt a little bit — and I've talked to women about this — where you sometimes feel you have to know *everything* before you can know one thing." She chalks the feeling up to the messages that women receive about their competence and how they perceive themselves. "We've noticed sometimes that when you would say a word to a man — we've been laughing about this in the office — you can use the word, and two minutes later he'd be using the word, maybe wrong, but it was *his word.* Oh, it was funny, we women would feel we had to know that word, be sure of that word, because we would be found out. There's that sort of impostor syndrome."

Charmaine's experiences are supported in research that examines the differences between male and female computer students. In a 2003 study, men and women who majored in computer science showed no disparity between ability, personality traits or comfort with computers. The only significant difference? Males perceived themselves as more competent; women perceived themselves as less competent.[24] Undoubtedly, this has an effect on career choices for each group. It affects how each group perceives the other. It also affects their willingness to ask questions and to meet challenges. One frequent theme in my discussions with women was the fear of looking stupid, especially in mixed-gender groups. This fear appeared to be a result of experience: male tech-users tended to be dismissive and negative towards users, especially females, whom they perceived as being less technical and as asking "dumb" questions. Lisa, a young project manager for a tech marketing company, was happy when she stumbled across an all-female technical community, because "it was a lot less intimidating and I knew I could ask silly questions and not be frowned upon, especially in a technology field when you don't know where to turn to get the answers. I'm not nearly as fearful now about asking a stupid question, but I think I needed to learn to be able to do that. I know now my questions aren't stupid. Now I wouldn't feel nearly as stupid asking a question to a man." Helen, though she is ten years older, agrees with this sentiment, and tells me, "I'm hesitant to show my greenness in front of a group of guys who can talk Web language in circles around me. I always get the sense there isn't that patience to explain something to me, whereas the female group, there usually seems to be the patience to explain things to people who don't know or who haven't learned at this point."

Tatiana, a graduate student who is interested in creating online communities for girls and women, is worried by this trend. She feels that attempts to encourage women aren't enough to create the fundamental changes needed in the climate of technological work. "What happens," she says rather gloomily, "is that we're encouraging each other but the men don't get it. The women working with the men don't get it, or they don't know how to get it." She tells me about an event that she went to, organized by a women's technology group. Normally women-only, the group had brought in a guest speaker from another organization that had both men and women, so men were welcome to attend the event. "I just happened to sit behind two men. When a woman asked a question that in their minds was a stupid question, they were snickering, and they were talking. In that respect, yeah, it should be women only, because that's why we're there, so we can ask questions and not feel stupid." The theme of women feeling stupid or undervalued in mixed-gender groups is recurrent and appears throughout the technical life cycle. Even young male users tend to overestimate their own abilities, and devalue those of girls, sometimes even to the point of physically pushing the girls away from the computers during classroom computer use.[25] When I speak to undergraduate students about this, they nod their heads knowingly. Angie, a young woman and self-confessed video game addict who is good enough to win money in video game competitions, tells me that since her early adolescence, she has been going to tournaments and communal video game playing events, and still many male players openly wonder why she is there. They harrumph that girls can't play video games and make it clear that she isn't welcome or valued. Fortunately for Angie, she has a thick skin, a tenacious spirit and, more importantly, a killer trigger finger.

Even when women do not underestimate their own skills, there are others waiting to do it for them. Lilith, a systems engineer who holds a very senior position, recounts an anecdote about bringing a junior level technician into an important meeting. She had started bringing these technicians into meetings because she felt it was important for the younger workers' career development to do some "job shadowing" of senior occupations. The job shadowing was intended to help the technicians learn what systems engineers do and get an idea of where they might need to develop their skills in order to be an SE. The junior technicians would just watch and listen, and not speak; their role was to observe and take notes. One day, Lilith had taken one of these junior male technicians along to a meeting. She was introduced to the client as the primary technical contact, and the technician was introduced as someone who would be observing as part of his job enrichment. After introductory discussions of services, the customer began to ask questions. "These were real questions, too," recalls Lilith, "about BGP routing and high availability networks. Our technician is good, but that's not his area of expertise." Yet the customer persisted in directing questions at the male technician. Lilith chuckles sardonically at the recollection. "The poor guy sat there wide-eyed, because he was put on the spot when he didn't expect to be, and understandably, he didn't know the answers." Each time the customer asked a question, it would be addressed to the technician, and Lilith would have to rescue him by answering for him. "Even after several answers," says Lilith, "when it should have been obvious that I was the one who was doing the responding, and that I knew what I was talking about, the customer kept asking the other guy, even after they were told that I was the expert, and I proved myself to be."

The third problem with human capital theory is its circularity, "since low levels of human capital appear to be both the cause and the consequence of women's changing participation in paid work during the life cycle ... Skill and productivity levels not only cause but result from relative wage levels."[26] Put crudely, nothing succeeds like success. Since lost wages, also known as the "mommy tax,"[27] are an integral part of at-home childrearing, a job which is primarily allocated to women, women may choose to acquire lower levels of training in advance of and during childrearing; this choice can be due to an informed evaluation based on women's lower wages. If women make a lower wage, fewer household earnings will be lost if women rather than their male partners leave the labour force during this period.[28] While this emphasis on the prescriptive and descriptive value of human capital theory is instructive, it may suggest that women are, in fact, less skilled than men. However, pay disparity between formally credentialed men and women is well substantiated.[29] Even when a man and a woman have the same degree or diploma, they are still likely to be paid differently. Thus, human capital theory doesn't differentiate between the *actual* level of human capital embodied in the individual, such as their formal educational attainment, and the amount of human capital that each person is *perceived* or *assumed* to possess.[30] Women's work is often assumed to be unskilled *whether or not it actually is,* and perceptions of skills are conflated in human capital theory with possession of actual skills.

This brings me to the fourth problem with human capital theory, namely, that it assumes that women (and, indeed, all people in low-waged work) are underskilled in general. However, an examination of women's educational achievements in Canada shows us that this is not the case.[31] Currently, women represent the majority of both university and community college students and recent graduates,[32] and their numbers have increased in post-secondary institutions dramatically.[33] Yet, even though girls are doing better than boys in school and women fill the hallways of universities and colleges, women are still much less visible at higher levels both within and outside academia, and women and men with the same credentials are still disparately compensated. Before they began to drop again in recent years, women's wages through the 1990s increased relative to men's, but this is in part because well-paying, traditionally male jobs have been disappearing or have become less available.[34]

The fifth problem with human capital theory is the assumption that males and females have equal experiences in and access to particular kinds of education and training. This is not necessarily so. While, as shown above, women are well represented in university undergraduate programs and post-secondary education, they tend to be under-represented in most applied science and technology fields — such as engineering and programming — which eventually command higher salaries.[35] Indeed, women's participation in these fields in universities has actually decreased since the 1980s. Applied science and technical institutions often have gender-specific gatekeeping functions that discourage women from participating or completing credentials in this field. If they do make it past the barriers, they can encounter numerous structural obstacles that range from an unfriendly climate to open hostility.[36]

While chatting with Stephanie, a nuclear engineer, I discover that the process of first-year hazing alone would be enough to eliminate many women. She attended the engineering program at Queen's University in the 1980s. In the early 1990s, Queen's was made (in)famous by the "No Means No" campaign. A campaign to prevent date rape at Queen's was answered by male students, particularly in engineering, with signs reading "No Means Harder," "No Means More Beer" and, most notably, "No Means Kick Her in the Teeth." Stephanie's hazing included crawling through vats of human waste and being on the receiving end of verbal and physical abuse. I have difficulty believing the scale of these events. It seems incredible to me. My first week in university as a fine arts student involved silly team-building activities, campus tours and trips to Canada's Wonderland. The most risqué thing that happened to me that week was getting free condoms from a safe-sex awareness group. I ask my husband, who went through engineering at McMaster in the early 1990s, if the intense hazing was common in engineering faculties. He looks abashed. He spent his frosh week pouring beer down his esophagus with hundreds of other manic males in organized drinking activities, and he can still recall all the words to the *Lady Godiva* song, plus another charming little school ditty that begins "Cocksucker, motherfucker, eat a bag of shit ..." If this is what happens on the first day of school at technical institutes of higher learning, no doubt many women are tempted to play hooky.

I am not arguing that engineering frosh week is the only reason for women's relatively lower enrollments in computer science or information technology (I also do not wish to conflate the cultures of engineering, computer science and IT, which can all be somewhat different). Being immersed in curse words and human effluvia during frosh week isn't necessarily enough to keep women out of technical education. The more important issue is the institutional mindset, of which hazing and gatekeeping rituals are a hyperbolic example. These events convey larger messages about the culture of many institutes of technical education in general, such as the creation of a certain kind of hierarchical community and the reiteration of gendered power relations.[37] Of course, not all universities and technical institutes have such dramatic, literal evidence of inequality. In most cases, the discouragement is not so explicit.

Nevertheless, women's under-representation in technical institutions in North America is a pervasive phenomenon. There are a range of intersecting factors that influence women's experiences of learning technology, and they operate on many levels. There are experiences of day-to-day discrimination and interpersonal difficulty. There are challenges around the gender composition of the classroom and what kinds of learners are valued. There are also issues of institutional culture. Women, who may have less money, reduced access to choice of training (such as women who are geographically isolated) and more demands on their time from domestic responsibilities, are often hesitant to enroll in programs that are expensive and of long duration (and, as a result, are less able to secure desired skills). In Canada, crucial federal and provincial support for women's entry into so-called non-traditional education has diminished in recent years.[38] Thus, some of the roots of occupational and industrial segregation, namely, disparate access to education and training, are laid down prior to employment.[39]

The sixth problem with human capital theory is that its grounding assertion — that increased human capital translates to higher wages — is false. A Statistics Canada overview of the 1990s labour market concludes that, in fact, during this period levels of education and experience rose while employment availability disintegrated.[40] Once we ascertain that workers have equal amounts of human capital, as assessed "objectively" by such things as certifications, then we should be able to see that they do indeed have equivalent experiences and earnings. However, this is not the case. In an examination of earnings patterns among Canadians with a recent bachelor's degree,[41] there was little difference found between male and female graduates in terms of skill level and preparedness for the occupations held, yet there was a significant gap in male and female earnings both immediately following graduation and in terms of increase over time (men tended to increase earnings much more quickly). Much of the gendered pay gap for recent graduates "has been associated with a general tendency for female graduates to have lower earnings than males within a given field ... regardless of the specific nature of their current job characteristics, post-graduation work experience, or personal attributes."[42] Studies of the IT industry appear to indicate that it is no exception.[43]

Despite persistent wage gaps, as I discussed in chapter 1, computer skills are nevertheless correlated positively with increased earnings. While improving skills attainment and access cannot close these gaps, it does have the potential to narrow them.[44] Women's access to training varies with the type of training offered. For example, studies show that women are most commonly trained in software applications. They receive approximately the same or slightly less professional/technical skills training than men, and less than men in other categories such as orientation on particular machines.[45] While there is a preponderance of evidence that shows a preferential male access to job-related training, especially employer-funded training, female disadvantages in skills training are likely related to occupational and industrial segregation, not gender dynamics within a particular job.[46]

Yet despite the general importance of computer skills for improving women's wages, there has been a *decrease* in wages for women that correlates with the introduction of computers into a workplace.[47] Often, as women acquire computer skills within typically female occupations, those skills become "feminized" and consequently devalued. Moreover, the implementation of IT into a traditionally female workplace has often been used as a strategy to reduce the number of workers required. Technology skill does not result in an occupational shift (i.e., moving women workers across occupations or industries) with consequent revaluation of women's wages; rather, technology implementation is occupation-specific, with remuneration given accordingly. In other words, female secretaries who learn how to make Web pages don't suddenly begin to charge $40 an hour for their work; they're still seen as "just" a secretary, if perhaps a more handy one to have around. Their knowing HTML may even increase the employers' expectations that they will take on more of a workload for the same pay. Indeed, even though men in sampled research groups had fewer computer skills overall than women, particularly word processing and statistical/analytical skills, they still earned more and had better chances of promotion.[48] The link between skill development and wages is highly gendered and, as a result, occupationally specific.[49] Training for women does not result in higher salaries despite its greater prevalence.[50]

Why, then, does human capital theory persist and remain so popular for explaining why some people don't do as well in the labour market as others? First, as I have said, human capital theory fits neatly into the mindset of corporate IT culture, which valorizes a consumerist free-market model in which innovation and information, not deliberate policy decisions, drive the economic climate. Second, it inhibits workers from proposing alternatives to the current economic arrangement. Third, it obscures the unequal process of skills acquisition — formal and informal education — and how these skills are valued, as well as how this might be stratified by gender, immigration status and other signifiers of social privilege. Finally, human capital theory does not critically examine the role that technology itself, or rather the implementation of workplace technologies, plays in *changing* people's skill sets.

Technology, Upskilling and Deskilling

Skills change. The value assigned to them changes. Both of these things are related to the role that technology plays in the workplace and in the nature of people's jobs. There are two common theories about technological change in the workplace that can be loosely termed "upskilling" versus "deskilling." The first theory proposes that technological change results in a more highly skilled workforce.[51] With the implementation of new technologies, workers become more extensively trained and demonstrate a more substantial skill base in order to utilize them. The second theory proposes that technological change leads to deskilling, or the downgrading of existing skills, and the eventual downsizing of the workforce (in other words, job loss and unemployment).[52]

While both of these positions have merit and substantive evidence, neither one adequately reflects the diverse experiences of technological workplace change. Both tend to divorce technology from its social, political and economic context, as well as from its situation in each individual workplace. One cannot discuss any process of skills change without asking "For whom?" and "How?" What does skills change mean, for example, to the auto worker? To the bank teller? To the retail service worker? Technological change in the workplace involves a variety of objects and practices, as well as diverse processes of development and implementation that are themselves informed by systemic dynamics (i.e., why and how these technologies are produced). Workers in one industry may experience technological change much differently than workers in another industry. There is diversity across industries, occupations, individual workplaces and across the range of social signifiers. The contention that job loss is a necessary result of the implementation of information technologies is at once too generalized and technologically determinist (not to mention gender-blind). It assigns a linear, definitive endpoint for an entire series of objects and practices. It assumes that the introduction of a computer is always done for the same reasons and always leads to the same ends. What is required is a more mediated perspective that takes account of the diverse material practices of technology as well as their intersecting ideologies, which both inform and conflict with their use. Technological change in the workplace does not affect male and female workers equally.[53] Industrial and occupational segregation, combined with structural issues in implementation, result in differential application of the new technologies themselves. A computer may be used to make a male draughtsman's job easier, while it may be used to downgrade or even eliminate a female clerical worker's position. Technology itself does not cause shifts in an organization nor is the use of any given technology an inherent feature, but neither do organizational dynamics alone explain technological objects and practices.

There are two main types of technological impact on people's work: alterations in the actual work practices of individual workers (for example, having to learn new software or use a new piece of equipment), and the job creation/elimination process that enacts large-scale institutional changes. Despite equal gender representation in workplaces undergoing general organizational change, women are more likely to be part of workplaces that experience downsizing as specific organizational change and are more likely to report a decrease in skills in these establishments. While technology is used to facilitate the performance of tasks, it is also a means for employers to alleviate staffing problems. The few employees who perform technologically facilitated tasks may have their skills increased in order to complete the new procedures, but those who are eliminated as a result will see their manual skills downgraded in terms of value.

Work is defined in terms of who does it. Since gender is a critical organizing component of work, technological developments within occupations are also incorporated along gendered lines. This is not to say that their introduction is straightforward, but rather that structures of the organization, which are arranged according to particular social relations, then mutually constitute technological development and implementation.[54] How work is done and by whom determines how technology is implemented and by whom.

How "skills" is defined is itself affected by power relations such as gender. Skills that women (especially particular groups of women) possess are more likely to be devalued no matter what they are, and investments in "human capital" are not equally remunerated for all workers. Skills associated with "femininity," also known as "soft skills," are the most devalued (despite their current promotion in the IT industry as valuable). But these gendered distinctions of skills are rarely mentioned when discussing technical upskilling or deskilling. For example, one skill set required of many female clerical workers is interpersonal relations: management of office staff, emotional labour in smooth maintenance of work and personal relationships, functioning as a liaison between levels of office hierarchy and so on. If these tasks are shifted to an increased load of information management skills such as data entry, resulting in greater social isolation, workers will often be said to have been upskilled despite a potential loss of this interpersonal skill development.[55]

We might even ask what is actually "known" by those who are seen to have certain skills. There is a difference, for example, between "knowing how" (i.e., understanding a process) and "knowing that" (i.e., possessing or retrieving information) and how these relate to definitions of skills.[56] We see a lot of discourse now about the "information age" and the "knowledge economy," but there is less attention given to the importance of knowing *how* to use the information of the age or what to do with the knowledge of the economy. In the course of my research, I spoke to many female IT managers who were frustrated with young IT college graduates. The managers perceived the college graduates as being less able to problem-solve and think critically. The grads bristled with ostensibly attractive commercial certifications, but seemed unable to broadly apply them.

Lilith tells me a story illustrating the problem with rote-trained IT personnel, whom she feels end up with very limited expertise. In one incident, she and a technician needed to perform a simple task using a UNIX terminal window. This window, commonly used to execute system commands, is a text-based window in which text scrolls from bottom to top. Commands can be issued to the system by typing in some kind of verb, such as "show," "run" or "config," accompanied by a specific thing that the user would like the computer to show, run or configure. It is similar to issuing simple commands that a small child can understand, such as "put on your hat." "More" is a terminal function that makes the page stop scrolling after all of the lines that are visible have been filled up. It allows the user to press a key to continue, so that the text doesn't scroll by before the user can read it. The technician was having difficulty following the instructions for this command, which were being given remotely by a superior. After trying several times to give the technician the command to turn off "more," Lilith suggested to the superior that the technician might find it easier if the instructions were in writing. The superior then e-mailed the instructions to the technician. The e-mail said:

> Please turn off more by issuing the following command:
> config cli more false
> Then send me the output of these commands:
> show config
> show tech

The technician read this e-mail and, at the command prompt, dutifully typed:

> P-l-e-a-s-e t-u-r-n o-f-f m-o-r-e b-y i-s-s-u-i-n-g t-h-e f-o-l-l …

"At this point," said Lilith, "we tried verbal correction, but to no avail. Finally we typed the command ourselves." The technician may have had decent computer skills, but lack of good verbal and reading skills rendered task performance impossible.

Ursula Franklin suggests that technological changes in the workplace have resulted in a work process where rather than each craftsperson being able to create, develop and finish a product using "holistic technologies," workers now use "prescriptive technologies," which break each job up into smaller tasks and organize the performance of work into small, manageable segments. Human labour is fragmented, divorced from the entire process, more easily watched and controlled, and the skill to develop something through a process is lost.[57] Skills that cannot be quantified or used directly for specific tasks are invisible. Instead of fostering innovation, suggests Franklin, such an organization of work constrains it.

While the learning of new skills might initially represent upskilling, in terms of new abilities acquired or the acquisition of new information, these new skills may not represent material rewards such as an increase in status. As one office worker remarks in Catherine Cassell's study on office workers, "What can you get promoted to? A bigger word processor?"[58] The increased expectation of credentials for low-waged, low-status IT work (such as being certified in various types of software to do data entry) means that marginalized groups of workers are in a double bind: to secure even the most low-status employment, they must acquire skills training that is often relatively expensive and time-consuming, yet that training does not translate into high wages but simply an entry into a dead-end job. For technological professionals, high-end training is very expensive (rendering it inaccessible to many individuals and smaller companies) and may not reflect the current skills nor the on-the-job experience needed. Companies may provide training to employees as a means of retaining their services and reducing turnover, requiring workers to sign contracts stating that they will not leave their current job for a period of time after taking skills training. Thus, training is not in and of itself a straightforward benefit to either employer or employee.

All of these factors can result in the replication of a skills deficit among particular groups of people. Access to skills training is unequal.

The ability to sustain ongoing skills training is unequal. And the valuation of the skills received is unequal. This adds up to a skills relationship that continues to privilege some people and disadvantage others. The technology plays a part, but not the primary role. Social relations of power and privilege shape who obtains skills, who is able to continue to upgrade and what skills are considered important and valuable.

THE CONVOLUTED PATH TO TECHNICAL SKILLS

Laurie, whom we met briefly in chapter 1, is a single mother in her late forties. She exudes moxie and can-do spirit. Her amiable personality fills the room with a sense of friendly, capable confidence. She's the kind of woman you'd want around if you were facing an intimidating task, such as smushing an icky bug, changing a tire, rewiring the house or battling Klingons. She'd not only get the job done right, but she'd teach you to do it fearlessly too. Within a few minutes of meeting Laurie, I can tell that despite her placid maternal appearance and cheerfully informal mannerisms, this is not a woman whose brain is content to lounge on the intellectual sofa. She has a diploma in electronics engineering technology, specializing in control systems. I confess my ignorance about this field to her, and she readily dumbs it down for me. "It's the precursor to robotics. That means things like programmable logic controllers, managing the machinery that is used in producing goods, like oil refineries, packaging and the automotive industry." Essentially, she makes the things that make other things.

I bob my head eagerly to show my understanding, and she continues on, lost in memories. "When I was nineteen, my friends were building guitars and stereos. I thought I wanted to do that, but I couldn't because I was a girl. My earlier work, we're talking thirty years ago, was data entry, because I was female. But when I was twenty-five, I got lucky enough to score a job putting in wiring for computerized cash registers. I strung wires from servers to cash registers." While I ponder the concept of being "lucky enough" to string wires, Laurie chuckles to herself. "I had to climb some very high ladders! In the Eaton's Centre, we had to go through the boiler room. I got to experience asbestos." She grins. I grin back. I cannot help myself. Her enthusiasm for technology is infectious. After a stint with Women in Trades and Technology, she was able to work in a lot of types of jobs not considered suitable for young ladies. She remembers, "We toured the generating plant for hydro, we worked on a car engine, I even made my own screwdrivers!" This inspired her to try out the electronics engineering program. She wrote entrance exams, got in and pulled off a 4.0 grade point average.

When Laurie graduated at age thirty-seven, she was the oldest person in her class. She dreamt of a future in robotics with General Motors. Instead, she landed an interview at Electronic Data Systems (EDS), excited by the fact that EDS offered training and programming opportunities. She was particularly looking forward to the chance to improve her programming skills. EDS had a surprise in store for her: they offered her a position as an administrative assistant. Laurie remembers, "I said, 'I've just graduated as an honours student, doing a three-year technical program in two and a half years!' But they hired me as a temp secretary." Shortly thereafter, she leaped at the chance to do electronic data interchange, or EDI. Though she landed the position, she wasn't given any work at first. "Even though I had a technical background, had my own PCs, had built electronic boards, this group didn't think that I could handle sitting on the phone dealing with customers and their technical issues with installing their EDI, whether it's network, software or hardware. I was solving problems, creating standards, doing data entry screen design, organizing user manuals for their helpdesk, chairing weekly technical meetings, but they still had this gender thing that because I had been an admin assistant, I couldn't be a techie. I sat for the first six weeks twiddling my thumbs. It was definite discrimination, and it was always there all the way through."

When Laurie looks back at her job history, she says she remembers being the only woman, or one of the only women, throughout. The only female programmer she can recall was a GUI specialist from Vietnam (GUI is techno-speak for graphical user interface). She thinks a bit more. In the last year of her electronics engineering program there were two other women besides herself. Both were skilled immigrants, and both could not find work. "One was an engineer from China, but they wouldn't let her be an engineer here. The other was from South America. Same thing." The help desk at the company had no women.

People who went to meet customers were not women. At EDS, she thinks, there was more representation, but still she found that it was male-dominated, with "more men in roles that would allow you to have authority, responsibility and decision-making capabilities."

I ask Laurie how she was able to slog through all the gender discrimination and persist in following her technological dreams. Her response is telling. "I grew up in a scientific environment. My father worked as a physical oceanographer and he took me to his work laboratories starting when I was five. He worked on underwater defense and they had a scientific research vessel. Daddy took me out on that all the time." Her childhood activities reflected an immersion in scientific and technical activities. Even her doodling paper was computer paper. "When you wrote a program in FORTRAN and ran it on the big computer, if you made a mistake, you would get all this wasted paper which was scrap paper. When I was young, my father would come home with this computer paper, perforated. Mum would take the short side to do grocery lists, then I would take the other side to do scribbles on it. My dad would program in FORTRAN and turn the radio to white noise to drown us out."

As Laurie speaks, I think of other women I interviewed. A connection clicks. Almost without exception, the women who are the most technically inclined and comfortable had parents, usually fathers, who supported and encouraged them. Regardless of formal education, parental support appeared to be one of the most significant factors determining whether a woman would develop and sustain an interest in and love for technology.

Remembering Laurie, I ask Lilith the systems engineer about her father. "My dad is a hard core electronics fan," she says. "He's an A/V nerd. He also knows how to do electrical work, which is why I have a complete lack of fear about working with 110V house current. He was totally into stereos and TVs. He had this thing in the seventies called a 'sonic hologram enhancer.'" I ask what this did, other than, presumably, enhance sonic holograms, whatever that is. Lilith isn't sure, exactly. She giggles as she says, "It had two buttons: 'on/off' and 'enhance.' When you pushed the 'enhance' button, he would say, 'Wow, listen to that!' but I couldn't tell that it did shit." We crack up together.

Other female techies have similar stories about parental influence on their early skills. Charity says, "My father is a big computer geek. I was using a Commodore 64, I was making little programs, when I was like, I'd be in grade four, so nine years old. Even before grade four actually. I grew up with computers. I played games and made little programs when I was young." Penny, an informatics specialist, experienced computer immersion at a young age, but unlike most of the women I interviewed, the influence came from her mother. "When I was five," she remembers, "my mom put me in computer camp. I can't think of anything nerdier. My mom was very progressive thinking, so every summer for about three to four years my sister and I would go to a computer-type camp. I did BASIC programming language when I was six." This foundation in computer skills led to enthusiastic participation in computing later on, but her interest in formal computing education as a route to a career was not strong. "I was always pretty good at computers, and when I was in high school, I took computers through high school, starting in grade ten. I never considered it a career because my guidance counsellor told me I would need to take calculus, algebra, geometry, physics ... I said forget it. That would be the most boring schedule. So that's why I didn't go to university. I've just been using computers all my life." Instead, Penny opted for self-instruction.

Erin's father was a psychiatrist who encouraged her to pursue scientific interests. When she was nine, she devoured his entire bookshelf. The reason was somewhat prosaic: it was a hot summer and the air conditioner was in his office. To entertain herself while enjoying the cool environment, she started at the left hand side of the bookshelf, and read it all the way through to the right. Father and daughter engaged in energetic debates as she read through his collection of scientific textbooks. Building on her youthful interest, she took computer classes in high school despite discouragement from guidance counsellors. Today, Erin is not in the least intimidated by technology. "It's not rocket science. The guy down the hall can do it." She relates a story of a friend in Saskatchewan who went to a computer class, which was an introduction to the Internet. "It was filled with all these Saskatchewan farm ladies in their polyester suits. They'd point to the combine and say, well, if I can drive that, this can't be any worse. And that's exactly that kind of attitude that really helps. I think a lot of what's important when you're starting is just an attitude of well, how hard can it be?"

*

The IT field is unlike many other fields in its typical path to education and skills acquisition. While there is increasing demand in IT for a university degree, it need not be computer science or a technical field.[59] Experience tends to be more valuable than formal credentials in most parts of the industry. Instead of deliberately planning a career in IT, people often arrive in IT by other routes, sometimes accidentally. Currently, IT work is marked by an interesting contradiction. On the one hand, the frequent lack of formal credentials means that people can access IT work through a variety of routes. Women who do not feel welcome in university computer science courses might acquire the skills and experience in other ways. They may sneak in through the back door by using the keys of other degrees, related work in another area, self-study or perhaps the mentorship of a more experienced person. On the other hand, the need for formal credentials is growing, as is the demand for particular skills and commercial certifications, which must be acquired through official and often expensive channels.[60] Formal education can be problematic for women, who are likely to encounter a variety of obstacles, particularly an instructional culture that is implicitly and explicitly associated with masculinity, and as such often unwelcoming and hostile to women. The IT field is becoming more codified and structured. To date, this tension between formal and informal routes to IT work has sometimes worked in women's favour. Will the back door eventually be padlocked?

Historically, the culture of IT has exhibited a whiff of anarchism and a quiet resistance to rules and regulations, including formalized "credentialization." Frequently, technical experts were made through hands-on experience, not born of the academy. In theory, anyone who had a talent for making 1s and 0s dance could be admitted to the IT club. This apparent lack of formal gatekeepers meant that technology could serve as the great equalizer, which rejected not only geographical and social boundaries but also systems of institutionalized educational power. Indeed, this situation might seem to provide a significant opportunity for women as IT users and workers, since women have traditionally been largely absent from computer science education.

While women seemed to be making inroads into the field in the 1970s and 1980s, over the last several years, women's computer science enrollments have dropped significantly.[61]

However, looking at computer science enrollment alone incorrectly assumes that a clear career path through formal post-secondary instruction in computer science straight to IT work is the norm. While computer science continues to define IT culture in certain significant ways, such a linear path is the exception rather than the rule for IT workers. Not a single woman I spoke to in the research for this book held a university computer science degree, though many had taken computer science and IT courses at universities, colleges and private institutions.[62] Indeed, there was substantial educational variation. Educational attainment ranged from some high school to a master's degree, and technological skills were gained through a range of formal and informal educational practices.

Some women did indeed develop an interest in technology as a result of enrolling in some kind of technical or scientific educational program. However, this interest was often discouraged formally, by the need for particular credentials, or informally, by practices of educational institutions. Erin, whose job was a very official-sounding resource system intranet facilitator, discussed the discouragement she had experienced about her choice to pursue technical education. After switching high schools in grade eleven, Erin drew up a program for herself that included biology, chemistry and physics. She met with the school guidance counsellor, who asked her, "Well dear, can you sew? I think you should drop physics and take sewing. It's very important to sew, and there's no Phys Ed in there. Why don't you drop Chemistry and take Phys Ed? This is too hard a program. I don't think you could handle this program." Recalling this, Erin laughs in disbelief, shaking her head. "I was coming in with an average of ninety-six percent or something, it's not like I had a fifty ... It was very weird. And this was in 1986." Obstructive guidance counsellors notwithstanding, Erin got the last laugh: she now holds a master's degree in science.

Even if they managed to get into a scientific or technical program, women confronted both overt and covert discrimination throughout their education. The IT culture developed in school and in the workplace often discouraged their involvement. They found a lack of recognition in formal education from instructors and exclusion from the students. Brenda describes her all-male computer class with characteristic frankness, illustrating the elements that might make IT education unattractive or unwelcoming to female students. "No offense to any of them, I'm sure they were all very fine people grown up, but you're talking geek central. Pocket protectors, the whole bit. Not a very sexy image. Not something your average seventeen-year-old girl looks at and goes, 'Ooh, wonder what that is.' If you weren't in, if you didn't know your COBOL from your FORTRAN, they didn't even bother talking to you. It was very cliquey, not very welcoming; nobody cared to explain to you what the cards were, or why they were full of holes." Erin concurs. As a teaching assistant while a grad student, she often had to comfort her students after they had been antagonized by the university technical help personnel while trying to set up their e-mail. "The tech people were *really mean*! They used to make my students cry!"

Preeti, now in her thirties, remembers her own undergraduate experience as a technical helpdesk worker, assisting students with a range of requests from how to renew library books on-line to how to install and manipulate UNIX. She found the climate of the technical department notably unfriendly during her tenure there. Bluntly, she says, "Male geeks are horrible, terrible, unhelpful, miserable, uptight people and incredibly misogynist." She was one of only two women, as well as a woman of colour. She felt that this contributed to the negative reception. "I'm bright and intelligent," she asserts candidly. I have to agree, since we have just finished discussing her thoughts on superstring theory, particle physics and non-linear time. Scientific and mathematic aptitude came easily to her, though she chose a liberal arts field of study. "But yet they saw me as not being able to handle the technical background. I never got promoted, and got constantly overlooked in favour of the males. The environment as a woman in that field was very trying. Guys didn't like women coming into their little group." While she was working to improve her skills, she says, along with her likelihood of getting promoted, her male co-workers were spending their time downloading pornography.

Both in research literature and in the perceptions of the women I interviewed, it seemed that many males who sought IT instruction tended themselves to be a socially beleaguered group.[63] Excluded from conventional norms of masculine superiority such as physical prowess, male IT aficionados often attempted to manufacture a new regime of masculinity that would include them. They created a culture that revalued particular elements of social interaction. They developed their own internal hierarchy that held technical skills in high esteem (building on a tradition of technological objects and practices being coded as male). Unfortunately, since many conventional definitions of masculinity are predicated on the negation or absence of women, female IT users often experienced overt or implicit rejection, regardless of whether or not they had actual skills. This pattern of behaviour extended into later experiences with the gendered IT culture of the workplace. Many women felt that despite their interest in and enthusiasm for the subject, active exclusion by male IT users significantly impeded their comfort with learning the skills.

Another obstacle to formal skills instruction can be cost. For example, while corporate training such as network certification is useful, it is prohibitively expensive. Lilith the systems engineer sighs wistfully as she envisions training possibilities. "I would love to complete certification programs by both Cisco and Nortel. Vendor training costs a lot of money, on the order of two thousand U.S. dollars a week, so you have to find a company who is willing to pay for that. And certification programs usually involve half a dozen or more weeks of training, so that's a pretty large investment for a company, and a prohibitive investment for a person. Small companies generally can't afford that. My company sure can't." As employers increasingly ask for these commercial certifications, people who cannot afford them are at a disadvantage.

Many women I spoke to had turned to outside institutions to upgrade their skills, such as community colleges or private technical schools. However, they had to bear much of the cost of this instruction. Some would have preferred to have other routes to this training, but others simply didn't bother wishing for it from their employer. Ann, who works in Internet marketing, sniffs dismissively at the idea that her employer might subsidize her training. "In terms of actual corporate requirements, the company I work for does a lot of ongoing, on-the-job training, but does very little to encourage (or pay for) continuing education outside the office. I take courses to expand my range of experience and capabilities. It has helped me evolve my position within the organization. They're not interested in coughing up some dough to help me out, but that's irrelevant." Luckily for Ann, she has the financial means to do so.

Not all women are so fortunate. A few women indicated that federal government-sponsored training programs had been fundamental to their skills development, and that they would not otherwise have been able to afford skills training. Debbie is a sweet-faced young woman in her twenties with a pink stripe in her blond hair. After high school, she had enrolled in university for a year and a half, but had to drop out because of illness. The long bout of unemployment that resulted from her health struggles and lack of credentials left her financially strapped and suffering from depression. By the time I speak to Debbie, however, she has a good job working as an intranet developer for a large technology corporation, thanks to government-sponsored skills training. She is very grateful to have been given the opportunity. "I know a lot of people who are like, well, people should be able to pull themselves up by their bootstraps, but y'know, I wouldn't have gotten off of welfare if it hadn't been for that government program that trained me to do what I do. I think it's way better for someone who's intelligent like me to be employed, rather than sitting there with five hundred dollars a month and doing nothing." I ask Debbie what might have happened if she had not been able to take advantage of this program. She thinks a bit. "I wouldn't be here. I'd probably still be doing retail, and not able to have a nice apartment and this kind of life at this point that I have."

Over sushi in a midtown restaurant, Sarah shares her experiences of unemployment and finding training opportunities. No shrinking violet, she is hardened around the edges, with frown lines carved into her face. She has been around the block a few times, and her cynicism permeates our conversation. She left her job as a Web graphic designer despite being successful at it. "I was sick of playing the corporate game," she shrugs. She doesn't like other women much either, despite her irritation when people read her as "butch." "I get told I'm extremely masculine. And I'm like, no, I just don't suffer fools lightly. Don't sit there and cry and tell me you can't do it, woe is me.

It's not working. I've seen women do that." I chew and nod silently, waiting for her to continue, trying to think of more non-threatening questions about gender roles in the workplace, hoping that she doesn't think I am one of those crybaby women she dislikes. Perhaps by way of excusing herself, Sarah describes her previous work climate. "You can well imagine starting working in my industry in 1990. I was always the only girl, but I was treated as one of the boys. I don't understand this playing the femininity card. I always got along with men." From her description, I have the feeling that being one of the boys involved some sacrifice on her part, but I am too intimidated to ask.

As her yarn spins out, and I listen to her complain about team building touchy-feeliness in the workplace, I also begin to wonder who had the problem. This was definitely a woman who was made for self-employment. Luckily for her, the opportunity eventually presented itself after she left her job and went on unemployment insurance. "When you're on UI, you get treated like you're dirt. They're kind of like, 'And your number is ...?' They make you feel like UI is equivalent to welfare." She was enrolled in a government-sponsored skills program that enabled her to start her own business and be funded for a year while she got off the ground. But, she says, "if I didn't have the backup and the money, knowing I didn't have to worry about the money for a year, I don't think I could have handled it."

*

Funding is crucial for skills training, and it is evident from the above examples that aside from skill valuation, the mere process of skills acquisition contains obstacles for those who are economically marginalized or whose employers cannot (or will not) support skills training. Support and resources for skills acquisition are thus critical for ensuring women's equity in the IT workforce. These formal and informal obstacles to the acquisition of skills resulted in a variety of responses: women forged ahead in conscious defiance, shifted into a more welcoming IT-related area of interest, used more "unofficial" forms of instruction such as mentorship or avoided technical education altogether. While most were successful at acquiring the skills they needed to perform their jobs, they were not always able to complete formal degrees and certifications or obtain necessary documentation.

Clearly, women can be successful in IT without possessing formal credentials. The non-linear educational path that appears to be the norm in IT work can sometimes work in their favour, as long as they can provide evidence of experience and hands-on competence. Given that women's experiences in applied technical instruction has historically been negative, it would seem that this quasi-chaos of credentials could be advantageous. Indeed, it would appear that IT work in general represents a radical reinvention of how skills and expertise have come to be defined. However, though it appears that a lack of formal instruction should present no obstacle to women in IT, and indeed that the seeming rejection of formal credentials might provide openings which other, more standardized professions might not provide, gender stratification has proven to be persistent and highly adaptable.

Like a virus, gender discrimination can easily mutate into other forms and infect whatever social system it contacts. In practice, the highly gendered, often rigidly hierarchical culture of computer science instruction often informs the culture of the IT workplace. In large part, the culture remains unabashedly masculinized, overtly or covertly unwelcoming to women, focused on hierarchies of mastery and display of particular skills. Technical skills themselves are associated not only with masculinity but also with power. Even if women do indeed possess the required skills, they might not be viewed as "technical enough" to adequately perform the work. Though what is valued as "technical enough" changes over time,[64] unequal binaries of gender continue to inform notions of whose skills are valuable and whose are not. Thus, despite an apparent rejection of stratification by formal credentials, the acquisition and use of skills remains invested with systems of power and privilege. We should bear this in mind as we watch women move into IT: Will we see more skills become "feminized" and hence devalued?

Currently, much attention is being given in Canada to the development of a skilled workforce, which includes an emphasis on credentials. This may represent a disadvantage to women who, as we have seen, are less likely to gain access to institutions and formal education in IT (despite their greater numbers in post-secondary education in general). In addition, skills that were considered rare and valuable five years ago may now be commonplace, and their value may have decreased. Indeed, while the expectation of credentials may continue to increase, their overall status appears to decline if they are associated with female-typed jobs, so that even low-status work may contain an expectation of extensive technical skills. The expense and difficulty of acquiring valuable IT credentials such as industry certifications, along with the frequent devaluation of the skills they already possess, may impede women's progress into senior positions in the IT workforce.

WHAT SKILLS ARE VALUED AND BY WHOM?

Preeti, whom we met above, talks a mile a minute, rambling over a wide conversational ground. She chats rapid-fire about her work as a DJ in the same breath as globalization, and follows it up with a treatise on philosophy. Unlike many of the women I spoke to, who came to IT from other "non-technical" disciplines, she started out in science and mathematics. Eventually, however, she switched into liberal arts because it more closely paralleled her understanding of technology and gave her a perspective on IT that she feels is richer and less constrained by scientific discipline. "What turned me on about the technology," she patters at me, "was that I was studying philosophy in school, radically rethinking my way of looking at the world, and getting into the physics-based theory of superstrings and particle physics. Superstrings seems to me more of a Buddhist philosophy: everything is connected in some way and there's a circular movement. I see the Internet as a network, totally dispersed, without a singular point, or one top pinnacle. I find that really exciting." I feel a sudden need for mind-altering pharmaceuticals just to keep up with her stream-of-consciousness thoughts. But weirdly, it all makes sense.

I think of Preeti during my interview with Erin, as Erin warms to her rant about the problem with the cultural conception of what "digital" involves. "Digital" annoys her, she says, because it implies one or zero. "All my work is in the gray area in between those two things. I don't like 'digital' as a model; I think it's misleading and inaccurate. The binariness of things is not really the way it goes. My work is so analog. I work with technology, but I also work with people, and my job, essentially, is to make the technology work in a way that the people want; I don't want to be focused on the technology, it's a tool for the people. The digital, putting the emphasis on the cut and dried, on the ones and zeros, is something that I really don't like." Erin conceives of her work as rhizomatic, interconnected, spanning the divides of human and machine. "I'm a bridge person. I can rip apart computers and put parts in if I have to, but it's not what I want to spend my life doing. What I want to spend my life doing is making the technology bend to people's needs."

Sadie Plant might agree with Preeti and Erin. In her book *Zeroes and Ones: Digital Women and the New Technoculture,* she argues that representing technology using the digital binary of 0 and 1 reproduces phallic symbolism. Zeroes and ones, she proposes, represent "Cartesian duality" of mind-body, male-female, with the spiky linear "1" symbolizing the phallic, skyward-thrusting male, while the "0" indicates the lack, the void, the chthonic chasm of the female. The female zero signifies nothingness, absence, and this has traditionally been women's representation in technoculture: nothing. Rather, Plant suggests, technology should be conceptualized as messier, interconnected, rather than discretely divided into binary couplets. This is not chaos but rather the subtlety of textiles: woven, knit, braided threads which meet at various points. It is a productive metaphor, says Plant, because such a network "has no governing point or central organization, neither subject nor object, only determinations, magnitudes, and dimensions that cannot increase in number without the multiplicity changing in nature."[65] Viruses, either the real or virtual kind, follow this model: their progress has no larger purpose other than to proliferate along lines of connection, host to host. Machines only seem to play by the rules. In practical fact, they are subversive, sneaky and, most importantly, constantly creating new possibilities.

Like Plant, Preeti has little patience for the traditional linear and logical model of technical interaction. Her view of technology too is fuzzy, netlike, interconnected, circular. "I've noticed that some people in coding, they cannot *think,* they cannot get their brains around the idea of hypertext, that everything's kind of *spherical,* they're still thinking really linearly. But that makes sense if you're coming from a programming background." This is, she feels, the advantage of a liberal arts background. Her interest is in applying the technology as a tool to facilitate communication, and this potential salvaged her concern with technology. "If I hadn't learned about the Internet as a communication medium after I rejected my interest in technology-oriented fields, I wouldn't really have looked at my computer as something that I'd be excited about. It wasn't until I became aware of applications on the human level that I came back to it, and I think that's the exciting thing about Internet technology."

According to Preeti, technology is revolutionizing the economy, in part because it allows people who have traditionally been excluded from technological practice to participate. "It's allowing people who haven't been very successful at regimented systems to have a home computer and think really radically and out of the box, and come up with third-party software and be Internet millionaires. The home computer is putting all this into the hands of the average person, giving us means to solve problems innovatively." I ask her what she thinks of the importance of a formal computing education. She is dismissive. "Universities have a reputation for being the places that grant credentials, but now employers just want to see a portfolio. Experience counts more than education. Universities don't know their own weakness, and the problem is that universities still overestimate their own importance."

*

Despite ongoing struggles over the value of technical skills and IT credentials, women working in IT have often resisted this gendered hierarchy of formal technical skills. Many women I spoke to who considered themselves quite technologically proficient were likely to view the technology itself as "no big deal"; they did not find it intimidating, gender-specific, nor overly challenging to master. They didn't feel that one had to possess any intrinsic aptitudes or abilities to use the technology, and they often noted that technological work was much easier than it is made out to be. While they recognized that different people had different technological needs and wants, they did not understand this as a result of gendered qualities. It was as if they had pulled the curtain back from the Wizard of Oz, and discovered that the ritualized, masculine magic of IT skills was mostly an illusion.

The women I spoke to indicated that it was sometimes the ostensibly "non-technical" elements that attracted them to technology use (though, in fact, so-called non-technical elements such as usability and communication are an integral part of the technology; they are simply not quite as valued). They felt that IT work tended to be represented as very limited and uninteresting, which stood in contrast to their diverse use of IT and the many opportunities they felt that the field offered them. They were generally confident about their skills as users, excited about the application of these skills to other areas of their work and personal lives and interested in ways to develop the technologies so that they would meet the needs and objectives of other diverse users. Many valued the non-technical elements of the field as much as (or even more than) the technical ones. Most did not regard technology as interesting for its own sake, though a few women did. Rather, they preferred to view IT as a tool that could be applied to a problem, or used to facilitate the development of new ideas, new ways of doing things and new projects.

In what is perhaps a direct challenge to the gendered value given to "hard" technical skills of machine mastery (the definition of which changes over time and according to shifting power relations), women I spoke to did not see being a "hard-core techie" as intrinsically valuable, or more valuable than having other skills. They tended to value "hard" skills as a means to an end; a way to help them find other, more interesting work. To some, formal computer science instruction was boring because they saw it as disconnected from pragmatic concerns and tasks, or limited only to theoretical concerns of programming. While all the women felt it was important to be able to use the technology well, and to be able to understand how it worked, they did not demonstrate an unequal binary evaluation of "hard" and "soft" skills. Their perspective on technology tended to be holistic, focusing on use and application rather than drawing on a pre-given set of valued attributes. For them, just knowing how to use technology was not in and of itself the most important quality, and they did not profess a hierarchy of valuation based on technological competence alone. "Technology itself I wouldn't go to bat for," says Erin. "The uses of technology are what I like." Though this is perhaps a false distinction on Erin's part, it illustrates the perceived and stereotypical dichotomy between object (male) and use (female). These women were indeed use-oriented and were critical of technology that was poorly designed.

They preferred to see problems with using technology as an inherent failing of the object or practice rather than a problem with the people who used it. This stands in stark contrast to much of the received wisdom about technological competence being an inherent quality of particular (gendered) users. These women argued that technical skills alone would not produce a good quality technological product and that interpersonal skills, not technical skills, were fundamental to the success of their IT businesses and careers.

Several respondents, like Erin, spoke of themselves as mediators, "translators," "bridges" or (to use a technical term) "interfaces" between technical and non-technical elements. They felt that they could use their diverse skill sets to act as facilitators between the often obscure objects and practices of "hard" technology and the needs of various users. Carmen, the information architect, tells me about how in her job she functions as a bridge between two kinds of people, speaking the language of each. "I found a place for myself on the technical team as the interface with the client, because my colleagues were more technically oriented than I was, so I found myself almost being a translator. This is not about technology, it's about communication. People that do best in it are people who've studied something to do with communication."

There was also the general feeling among the women that IT was not a field that should be restricted to certain groups or to certain disciplines. They were proud of their ability to combine various interests and aptitudes. Many felt that the technology was an entry point for the creation and development of new kinds of objects, practices and conceptual structures. They did not mention logic, linear thinking, rationality or other qualities traditionally associated with technology use. Instead, they spoke enthusiastically about things like problem solving, communication and artistic impression. In fact, some felt that an exclusive focus on the acquisition of formal technical skills could be an impediment to new ideas and developments in the field. Carmen saw her role as the "prodder" of the IT department; she continually challenged her staff to implement new kinds of solutions that met the individual needs of the client (perhaps in unique or innovative ways). "I try to keep my eyes open for where I've seen other things work, or where I hear that there's new technology or new information that will make a better user experience. Technical people may want to use the existing solution they're familiar with, rather than going into new territory, and sometimes we have to help urge them along." The potential these women saw in technology was extensive, and they expressed their excitement about being able to participate in technology's development and evolution. They did not link their success in IT exclusively to their possession of particular types of skills.

The emergence of the Web as a viable communication, information and artistic medium has resulted in an enormous expansion of possibilities for use, development and creation. Many women, particularly those working in jobs which were "interdisciplinary," stress that it is this combination of the technical and "non-technical" elements that make work in this field appealing. "It's a nice blend of technical and creative," says Anya, a Web designer. She feels that this combination was part of the reason why the Web has exploded, because people "were looking for something that combined the disciplines in a particular way, and there really isn't anything else that does that immediately at a lower level. People find the work very interesting." Jamie, a young technical writer, is excited about the possibilities for melding her educational skills with her IT skills. "In ten years I want to be doing instructional design and computer-based or Web-based training. I think it's just a fascinating field, and there's so many possibilities for really neat things. Either structured learning through your computer or designing a learning experience that people can have through their computer … Something that really has the learner's needs in mind rather than the teacher's needs ... I think there's definitely a place for both in-person and technological learning."

Clearly, there are innumerable "non-technical" interests and fields of expertise that could be facilitated and improved by these women's involvement with IT: writing, marketing, artistic production, project management, publishing, health care and so on. They view "hard skills" as a means to an end rather than an end in itself. Directly relevant skills, particularly technical skills, while important for performing tasks, are not the sole determinant of job performance or work quality. This stands in stark contrast to a discourse that suggests that so-called hard skills are required in order to do good work in IT.

*

This chapter has sought to challenge the notion that skills are a straightforward matter that can be viewed as commodities that one may purchase to improve one's work value. Rather, the process of obtaining skills is not solely an issue of opportunity, and what counts as skills is open to negotiation along axes of power and privilege. In the case of IT work, skills acquisition alone is not always sufficient to perform well or to use IT in innovative, productive and practical ways. Yet the often deceptive concept of individual skills acquisition as a direct route to status and compensation persists. This places a double burden on the IT worker who is having problems with the acquisition and valuation of her skills: she is both impeded in her attempts to obtain formal credentials by structural relations that are made invisible through discourses of individual empowerment and may be punished for not having such skills, whether or not that is actually the case.

Over ten years ago, Ursula Franklin argued in her work *The Real World of Technology* that technology had come to be defined as a way of doing something, and that technological practice required defining a knowledge community (in other words, a group of people who have access to ways of knowing and doing).[66] "Who knows?" became a statement of social power. Communities of technological practice produce, and are produced by, particular relations of gender. Certain kinds of knowledge are thought to be the province of masculinity or femininity, and "border crossing" is both economically and socially discouraged. This knowledge community, if it wishes to clearly define its boundaries, institutionalizes a "credentialling" of expertise, so that only certain groups of people are able to use or control the technology in question. Those who are already inside the community can serve as gatekeepers in deciding which new learners can gain access. The proliferation of private technical education suggests that while more institutional bodies are interested in providing credentials, they may in fact be only raising the bar for which skills are defined as "hard skills" and, consequently, for determining which skills are valued enough to receive substantial compensation. Thus, while it appears that many more people may be admitted to the community of knowers, the inner circle of knowers reinforces its boundaries, and unequal relations of power and privilege are replicated.

As I have stated, IT work is positioned at an intersection between a resistance to formalized skills and an increasing reliance on credentials. Franklin's concern was that work and knowledge would become increasingly fragmented, so that holistic skills development, similar to a medieval craftsperson who could oversee a process from creation to completion, would become less possible.[67] Interestingly, a recent MIT report suggests that the community of knowers is returning to a model like that of medieval knowledge guilds. Making a career in IT, the authors suggest, "now involves progressing through a series of assignments that provide continual opportunities to learn. In many situations this represents a return to a craft mentality, where progress is not represented by position, but by growing mastery."[68] History tells us that while women were indeed active in medieval guilds, they were also often explicitly excluded from them. Later, other forms of skill communities such as trade unions were openly hostile to the participation of women and immigrants. Does the call for skilled labour represent more opportunities for formerly marginalized groups of IT workers or merely a redefinition of knowledge boundaries?

In the next chapter I examine more closely the process of acquiring IT work through certain types of instruction and the promises that accompany this process. I explore the apparent contradiction between opportunity and access and the possibility of increased structural gatekeeping within the workforce, particularly as it relates to women, and I ask many of the same questions I asked in this chapter. The relations of skills are just that: relations. Skills are not a thing or a commodity, but an expression of a social relationship.

CHAPTER 3

Great Promises versus Material Realities

Tari and I arrange to meet at a coffee shop in the east end of Toronto. She tells me how to find her: just look for the short hair and snowboarding jacket. With these visual cues in mind, I imagine a punky urban skateboarder kid with spiky Technicolour hair. Instead, I wander into the coffee shop to discover said snowboarding jacket being worn by an amiable, bespectacled woman approaching middle age. "Oh yeah," she grins mischievously at me, "I forgot to mention I was black." I like her already. It's hard not to. She is a ball of energy and an engaging, articulate storyteller. She quickly gets me giggling by narrating, with perfect comedic timing, an anecdote about how she recently broke her nose by sprinting face-first into a glass panel in her office. Despite the occasional workplace pratfall, Tari is an experienced Web developer with a background in business administration and has started many of her own companies. Resilient after what she calls the "dot-bomb," the downturn of the IT sector in which she lost her job, she formed a company with two programmers. The formation of this company echoes the practical ethos of many startups: the three of them ordered in chicken wings, fired up the computers and brainstormed business ideas late into the evening. Once the business was up and running, the versatile Tari handled customer service, project management, acted as liaison between the programmers and clients and even did some of the programming herself. In 2001, she became interested in hand-held computing and put up a Web site that provides information and resources. Her enthusiasm is unmistakable. She seems to inhale knowledge and exhale ideas. As she puts it, her entire career began with a twenty-dollar course on HTML she took several years ago, which set her on the path to gaining a broad range of expertise. By the end of this course, she could look at lines of code and immediately see mistakes, and she was soon helping other people with their programming. Tari has a good instinct for what new technologies will be important or hip, and is skilled enough in a range of technical disciplines to teach and develop various courses at community colleges and private institutions.

But Tari is currently working at a call centre making eleven dollars an hour. This, she tells me, is a "new refuge" for downsized tech workers like herself. Many combine call-centre work with contract self-employment or multiple jobs, or simply put in long hours at the centre to make ends meet. According to her, "people now are lying, downgrading their résumés to find work, because they're overqualified." In the call centre where she works, she says, there are many people with four-year university degrees. I ask Tari to describe the demographics of the centre. "There are about three hundred or two-fifty non-white employees," she says. "The black Africans are highly educated in high tech, while the black Caribbeans tend to have only high school. There are single mothers with kids. They want smart people who speak English without an accent, but they don't want people who are rebellious." Middle management at this particular company, reports Tari, is of mixed ethnicity, but upper management is almost exclusively white.

I ask Tari if I can talk to someone else that she works with. She hooks me up with Mike, a man in his thirties at the same job site. What is interesting about Mike is that he is white, male, young and well-educated, precisely the demographic that has traditionally done well in IT. He had a technical father who was a systems analyst, a factor that I identified in chapter 2 as an important one in assuring IT career interest. He has a bachelor's degree and job experience working in Web design, computer animation and the Y2K initiative.[1] From his description, it also appears that young men are finding their jobs increasingly downsized and eliminated, ending up in job centres formerly reserved for traditionally marginalized workers. "Things got bad after Y2K," Mike remembers. Hired in a frenzy of worried activity, in preparation for an anticipated millennial calamity, Mike lost his job after January 2000 came and went without so much as a server fart. His IT position in a non-profit organization was eliminated. In an effort to expand his skill set, Mike moved to Montreal to learn more of the "creative side of IT," such as Flash animation. Still, he couldn't land a job in the field. Eventually, when his Employment Insurance ran out, he ended up in a call centre and has now worked in call centres in both Montreal and Toronto. Asked to describe the call-centre workforce, Mike identifies differences in the two locations. "Toronto is minority based, single moms, young guys. Then there's the transitionals, the people who've been there, done that, EI's run out. There's about ten percent of that. In Montreal it was much bigger, it was primarily Anglos trapped in the tech sector after the bottom dropped out and the money disappeared. There's more education and not as many people supporting families on it."

Many of his Toronto co-workers are clearly underemployed. Mike describes the co-workers who have a background similar to his. "One person is a database admin, trained in a private college ... another trained in a private college for computer animation and design. Web administrators, server administrators, there are a few of those. There are people who were doing IT contract work who got tired of being screwed over when things got bad." Training for this job ranges in duration, but according to Mike, training modules are developed for a user with approximately a grade seven education. In preparation for his job in Toronto, Mike took "two days of customer service training, such as when to say yes, when to say no. Product knowledge is about seventy percent of that. To apply for a job all you need is a pulse and a good voice, to be polite. It's really more of a smell test. The most recent training was two weeks, but still, it's not rocket science."

Mike is glum about the conditions of his work. He describes a Dickensian workhouse environment that glorifies the rhetoric of mass production while employing the techniques of a globalized sweatshop: contingent labour, worker surveillance, interchangeable "flexible" labour. "It's a factory, it's giving people a different feel, a production line feel. They use production terms: the shop floor, the clients, a lot of acronyms. It gives it a false sense of importance. It's all about spinning you around, demoralizing you, you're always on the verge of losing your job. Orwellian monitoring, bottom-line management, numbers and figures and turnaround. Personality and anything like that is gone. It's very modular, very cubical, no matter which campaign you've been trained on, you can be pulled from one and put on another." Advancement in this field is limited, and workers often move from one call centre to another. "All those things you took for granted that a job gave you are gone — job security, references; you can get the mobility but there's nowhere to go." In terms of management, he says, of course "you'd much rather be a peon than responsible for peons, you're gonna take a buck more an hour to watch other people do your job." He describes workers who have stayed longer than eight months as "lifers." Basically, he tells me, call-centre work consists of "selling shit to other people who are doing shitty jobs. You don't have like a drawer, it's your bag, you're nothing more than a temp worker … It's kind of like a feedlot, you chew your cud and you leave."

I am reminded of Tari and Mike when I read a Human Resources Development Canada report on call centres, which states: "Ontario is an attractive province for call centre establishment as over half the province's workforce has some post-secondary training."[2]

*

The early days of the so-called dot-com boom were heady times for employers and employees alike. Old rules, from dress codes to business strategies to age-dependent workplace hierarchies, were being challenged. The new rules were simple: there were no new rules beyond the inevitable march of progress and the expression of individuality and freedom through creation, use and consumption of technology. In his wonderfully titled book *Cyber-Marx,* Nick Dyer-Witheford outlines what he views as the central assertions of the "information revolutionaries" of the period.[3] The first assertion is that the world is undergoing fundamental change, with technology as the agent responsible. Second, that the most critical natural resource is "technoscientific knowledge" and people who have this knowledge will form a valuable upper layer of society. Third, that wealth could somehow be generated from interaction with and representation of symbolic data and information. Information alone was a valuable enough commodity that its exchange alone could create profit for those who own it. Fourth, that these technologically inspired changes would necessarily be good and beneficial for all. As Dyer-Witheford sardonically notes, "A brilliant culture of individual and collective self-actualization is seen arising from the matrix of the networks."[4] Any problems with this global technocultural phenomenon would be like glitches in software: eventually, with an upgrade to Society 2.0 and a debugging — which would rid us of antiquated concerns like how to put actual, edible food on the table — our system would run smoothly. Finally, this technocultural revolution that could bring wealth, prosperity and a collective finding of enlightened headspace to everyone would, at its end point, result in something like the creation of a new species, a fundamental reinvention of human consciousness; an existential reboot, if you will.

In the early 1990s these promises appeared poised to come true, at least for a few lucky folks in Silicon Valley (unfortunately, as the concurrent Justice for Janitors campaign showed, these promises did not appear to apply to the vast numbers of support workers who provided the invisible greasing of the technological machines like Apple, Intel, Hewlett Packard, Oracle and IBM).[5] Millionaires and geniuses emerged like Athena, full grown, from their basements and dinosaur corporations. Hyperbolically gendered terms were (and still are) used to describe them.[6] A quick glance at IT industry publication *Fast Company* reveals use of "mogul," "guru" and "evangelical business revolutionary," while *Wired* enthuses over "fearless leader," "founding fathers," "masters of the universe," "boy wonder," "empire builder," "visionnaire" and "atomic rulers of the world." In the breathless accounts of 1990s technoculture, the iconic "Technological Man" smashed the emasculating stereotype of the four-eyed, pencil-necked dweeb: he was an explorer, a freedom fighter and a libertarian messiah. He rode through the electronic frontier like a conquering hero. Grandiose narratives proliferated. Former recluses for whom technological competence existed in an inverse relationship with social skills self-invented themselves as macho supermen before whom Nietzsche would tremble. Women were absent from this vision, except perhaps as damsels in distress providing the goal of a video game; nice ladies who needed protection from naughty words; porn stars with digitally amplified pulchritude; simpering "booth bunnies" at IT trade shows; or scantily clad virtual babes who clung adoringly to the pant leg of heroic supergeeks. As a decades-old children's book explains, "boys invent things, and girls use things that boys invent."[7]

Given the phallic significance of a big wallet in the "New Economy," one of the most seductive promises of the early days of the industry was the possibility of making enormous amounts of money. A small population of nerds and geeks could command salaries that their parents could barely have dreamed of, merely because they had a fondness for tinkering in their basement or could chat fluently with machines. The ideal form of work practice, according to mainstream publications of technoculture, was for highly skilled professionals to sell their intellectual labour power to a deserving employer. Employers, in turn, would offer copious compensation and the possibility of lucrative retirement at age twenty-five when the dot-com went public and employees cashed in their stock options. Survivors of the dot-com crash are more cynical now. "In the 1990s, people just lost their minds," Andrea, a dot-com survivor, tells me earnestly. Where she lived and worked in California's Bay Area, "banks were taking stock options as real estate equity. People were accepting ridiculously low salaries because they were promised stock options." While salaries for professional and semi-professional work in the IT industry continue to hover well above the national average, the money bubble appears to have burst in the past few years. Andrea is unambiguous in her advice to newbies: "I tell people now, 'Ask for the real money, not the fake stuff. Get a good salary, not stock options. Stock options are imaginary money.'"

Tari is also characteristically pragmatic about the employment situation. "There was a lot of bad money funding stupid ideas. There were bad business plans, and people burned through money. In the Web industry, there were lots of people with quickie educations in high positions who were overpaid and had high expectations. And their companies sucked." During this period, employed by various IT companies, Tari got herself into hot water. Trained originally in business administration, she felt the need to point out to her superiors when she thought they were making poor decisions. "The fundamental rules of business still apply," she says, gesturing for emphasis. Profit cannot be generated from nowhere; companies needed a solid business plan. "This isn't new business," she exclaims, "this is the same as it's always been for ten thousand years!" This frankness in critique caused, as she puts it euphemistically, "interpersonal conflict." Her situation was not helped by the fact that she was outspoken, female and black in an industry which, she feels, despite its relative youth, still plays by the rules of the old boys' club. In the larger companies she worked for, she says, she found an "old school way of thinking," which was not receptive to this kind of feedback. When pointing out what she thought were the problems in a business plan, she was told that she was negative and not a team player. She remains critical of the mindset of the period. "Companies, drunk and stupid on the so-called Internet age, hired people and gave them ridiculous titles and offered them stock options, most of which turned out to be worthless, and even turned on them, creating nightmare tax situations." Luckily for Tari, her business background and pragmatic nature stood her in good stead when she refused to take stock options in lieu of compensation. "I took cash instead."

CHANGING WORK PRACTICES

Changes in work practices do not happen in isolation, and there are many elements to them. There may be changes in how work is conceptualized, in the structural organization of a workplace, to the process of work and in how workers are understood and treated by the employer, both on an individual and a structural-institutional level.[8] Each of these are interdependent and vary from workplace to workplace, and from industry to industry. In terms of IT work, there are several reasons for the shift in work over time, and the reasons incorporate all of these factors.

First, many early IT companies, while they represented a challenge to how work was conceptualized in that they were likely to be small, upstart firms with a flexible view of work, did not (as Tari notes) have a solid business model. Simply speaking, they were unable to figure out how to make a sustained income from the new technologies. Some appeared to have hoped that if they built it, customers and users would come. Technological literature positioned the IT manager as a heroic "visioneer," an agent of revolutionary and productive technical change.[9] He (for it always seemed to be he) would part the antiquated, analog sea and lead his people to the promised land of IT gadgetry for all. Companies invested heavily in the production of expensive products and systems of infrastructure that were never sold. Many accounts of sales and profits were deliberately fictionalized and "massaged" to live up to the market hype. Numerous scandals emerged, such as the WorldCom fraud, Lucent's firing of several executives for violations of the U.S. *Foreign Corrupt Practices Act* and Nortel's admission (after laying off 60,000 workers while top executives collected bonuses as high as $2 million) that its accounts contained hundreds of millions of dollars in "irregularities" as far back as 2001. Eventually, capital investment in the field as a whole declined, business models did not prove their worth, financial alchemists failed to find a way to turn information into gold and many firms went under.

Second, industrial organization has changed. Most of the utopian predictions about the information age made in the late twentieth century tended to view the idealized career trajectories of white male senior level IT professionals as the norm. These predictions saw work as liberating, creative, meaningful and always under the control of the worker. Their vision of the "information revolution" involved an affluent "knowledge class," a technobourgeoisie of educated professional workers retreating comfortably into their electronic cottage. The rote labour of the worker in the microchip factory, the data entry clerk or the deskilled machinist did not have a cameo in these epic stories of technological progress; neither did ecological destruction, technological fallibility or economic downturns. Nor did these narratives allow for the possibility that the idealized white professional male might also lose his job. In discussing this trend, Dyer-Witheford is blunt:

> Analysis that sees "symbolic analysts" as the crucial actors in globalization does not grasp the speed with which capital tosses yuppies from the lifeboat when cheaper replacements can be found. Even symbolic analysts feel the blast of globalization, as North American computer programmers are undercut by Lithuanian or Indian competition, and architects, engineers, and professors discover that those who can telecommute can always be teleterminated by cheaper services uploaded from anywhere on the planet.[10]

In defiance of the notion of infinite upward progress as a result of scientific expertise and the proper gadgetry, many sectors of the

IT industry, such as telecommunications, experienced spectacular explosions followed by likewise noteworthy implosions. In general industry terms, traditionally male-dominated, unionized goods-producing sectors underwent a decline relative to the traditionally female-dominated, non-unionized service sector. Jobs in feminized service-type work, as I noted in chapter 1, tend to be precarious, lack benefits and security, have lower wages and frequently involve temporary or contingent arrangements. The IT field has not been immune to this industrial trend. Currently, service-type industries in the IT field continue to grow, while goods-producing industries appear to have slowed their increase.[11] And many of the remaining IT jobs increasingly resemble feminized service-type work in their organization. This shift to a service-based IT industry may be one explanation for women's growing numbers in IT.

The third element related to changes in work practice is the change in North American labour and production patterns that has come about as a result of globalization and transnational capital investment. Corporations are now able to move their resources more freely across national boundaries, especially with the assistance of technology. They can locate their head office in one country, their production site in a second country and their distribution centre in a third. In theory, they can look all over the globe for the best place to meet their business needs. This has resulted in shifts in work practices and remuneration. For example, some types of IT work are now "outsourced," or contracted out, to areas where the cost of labour is cheaper. The advantage of the virtual globalized workplace is that it is as easy to send it to the east coast of Canada as it is to India or the Caribbean. This arrangement can mean that the employer can shop around for the lowest labour costs and the least amount of labour legislation and regulation, and can hire and fire employees and subcontractors as necessary. For women in poorer regions of the globe, this means that they are often in demand as workers in this new economic arrangement, since they are seen as diligent, "nimble-fingered" and, above all, cheap. Depending on the type of technical work required, and the cultural context, it can also be seen as a socially appropriate extension of existing women's labour. For example, data entry workers in the Barbados, as described by Carla

Freeman, are given a fairly high status as "office girls" who are smartly attired and enjoy the privileges of an air-conditioned office, unlike their sisters labouring in the fields.[12]

This type of libertarian global free market model is in keeping with the view of IT employers as techno-cowboys, setting out to conquer the frontier, rejecting the limitations and boundaries of the "old ways" (the old ways, of course, include such antediluvian issues as employee protections and employment laws, and these are held to be antithetical to progress and innovation). Such practices are justified by the notion that the ideal virtual organization is really composed of individual, autonomous agents who are able to sell their skills to the highest bidder and that IT workers prefer the freedom of flitting from job to job.[13] In the early days of the Internet, the metaphor of the Wild West was applied to indicate that the Net was an anarchic place where a man could go and, with enough moxie and entrepreneurship, find a homestead to hang his hat. It was an escapist fantasy enjoyed by many white middle-class men whose actual experience of a Western frontier was limited to eating at Taco Bell. Women, as usual, were largely absent from this vision. Nice ladies, it was assumed, did not participate as trailblazers and needed to be protected from the unkinder elements.[14] Like other fantasies of the Western frontier, it is also notable for its amnesia about who and what were actually present before marauding settlers arrived and claimed the terrain as their own. It is a myth that is imbued with notions of the moral rightness of territorial domination, and particular ideas about what constitutes freedom (freedom-from rather than freedom-to). These days, the colonialist elements of the Wild West metaphor are becoming apparent in global IT labour arrangements: there are only a few powerful hired guns, the treaties are inequitable for indigenous people and the land is littered with a whole lot of dead buffalo.

The fourth element that may change work practices is the emergence of private IT instruction and certification, which aims to produce more qualified IT workers but which may in fact eventually result in an increase in officially credentialed but low-status work. While reports still identify a shortfall in recent graduates in terms of industry needs, the proliferation of private and public IT instruction has resulted in a larger pool of labour from which to draw. In addition, the expectations of many jobs for formal credentials and commercial certifications have increased. Nevertheless, as I noted in chapter 2, IT instruction and certification is explicitly linked to better wages, career paths and working conditions. Graduates of private IT instructing are increasingly likely to emerge from their studies into a job market where their abilities are devalued, their work process fragmented and downgraded and their occupations feminized in a consumption-oriented, service-based economy.

A troubling trend for those of us who work in the field of academia is the unambiguous link made between education as career preparation. This career preparation is not envisioned by academic marketers and policy-makers as a general readiness for responsible social, political and economic citizenship — which perhaps include a broad base of skills in problem solving, critical thinking and research — but as instruction geared towards creating very specific kinds of workers, in a specific kind of way. University funding is increasingly supporting technoscientific projects that are of immediate benefit to private enterprise. The pace of instruction is sped up and quickie certifications proliferate. What employers demand is dictating educational content. Workers are told that they must continually upgrade their skills in order to remain marketable commodities. However, the economic cost of acquiring skills is increasing, while the return on investment is not guaranteed. Women, in particular, are vulnerable to this message since, as I discussed in chapter 2, they are already viewed as less skilled. Doing a short certification course at a private IT college may seem like just the thing to help many women with their career advancement or assist them in breaking out of a pink-collar job ghetto. When they emerge with a certificate, however, many discover that IT companies are unimpressed by their achievement, viewing it as largely worthless and the product of a diploma mill. They also find that the unemployment line is crowded with other people clutching similar pieces of paper. They may wonder if they received the right kind of instruction. They may wonder if there really is a skills shortage in the IT sector.

In the remainder of this chapter, I explore two of these trends: the shift to service work and the role of IT instruction, and the implications of both of these trends for women's work in IT. I detail some of the changes in work organization that affect women's IT work practices and experiences and suggest that the promise of lots of lucrative IT work to go around is challenged by the material realities setting in to the IT sector. The IT sector is currently experiencing a downturn, and while many women are indeed gainfully employed, how long might this last? Shifts in work will likely mean that work will remain precarious, non-standard and casualized. This has some sobering implications. While, as we have seen, skills acquisition is viewed as important, the relative ease of enrolling in private IT instruction may flood the market with qualified candidates, driving the price of labour down and the competition for good jobs up. There is a strong chance that the sediment of the IT field will "settle" out into a thin upper layer of well-paid workers and a thick lower layer of poorly paid ones (and, in the course of this process, the usual divisions along race and gender lines will be maintained).

The Rise of IT Service Work

Service work is characterized, in part, by its preference for precarious work practices. Service workers are more likely to be low-paid and considered low-skilled, and less likely to be protected by unions or workers' associations. With the diminished presence of worker protections, job security and benefits are largely absent. In many service industries, labour turnover is high. Service work continues to be the predominant work performed by Canadians.[15] In general service work is more likely to be performed by women — in fields such as health care, social services and retail sales — and be associated with the stereotypical feminine characteristics of caring, selflessness, servility, maternalism and decorativeness. Immigrant women, in particular, are disproportionately found in service work, and they have become associated with this type of labour.

Although the ideal service worker is, to some degree, invisible, at the same time she is often under constant surveillance. Technology is often used in service fields to monitor and quantify all aspects of work. For example, computer software in restaurants tracks what food is served and how rapidly. Displays that pop up on cash registers prompt fast-food cashiers to ask customers if they would like to supersize their order. Retail salespeople sport little microphone headsets in order to respond swiftly to the demands of their supervisors. And yet, though employers may attempt to regulate service work using assembly-line principles, the nature of service work itself resists this discipline.[16] Much of service work is difficult to reduce to numeric units of production. Work that involves the development and maintenance of interpersonal relationships or that continually needs to be performed (rather than completing a product), as service work often does, is hard to quantify. While we know that a nurse, or a store manager or a child-care worker must have certain formal qualifications before she is allowed to practise her profession, a significant part of her job involves skills and the execution of tasks that are immeasurable. A child-care worker, for example, knows that all her charges are not alike: some children may need extra attention, they do not follow preplanned schedules neatly, they may throw up or shove a marble up their nose at the least opportune time. The child-care worker depends heavily on experience, interpersonal skills and intuition to do her job. She expects the unexpected. Her job defies routinization. Child development does not occur in a precise, predictable, easily measurable way, and thus her service and caring work cannot be measured by the clockwork standards of a factory.

It is sometimes assumed that service workers are less skilled than workers in the manufacturing sector, and in chapter 2 I suggested reasons why this might be so, including the fact that the face of service work is less likely to be white, middle class and male.[17] I will add now that service work is considered less skilled because sometimes the skills involved can be hard to observe and measure. In the case of many female service workers, it is also considered less skilled because the skills associated with it are seen as a common-sense extension of the feminine role, nurturing rather than educational or economic.

*

Proponents of service work emphasize its role as a rapid-response strategy to shifts in global production. A document produced by Human Resources Development Canada chirps merrily, "To remain competitive, companies these days need to be ready to engage a customer at any time of the day or night and, in a global economy, anywhere in the world."[18] Like a team of trained commandos, service workers are deployed when necessary, airdropped into hostile territory, then left to fight their way out of the jungles of the transnational market. They are agile, their skills are flexible and they respond rapidly to shifts in demand with razor-sharp reflexes. They can work long hours without sustenance, with the well-being of the parent company their only concern, and their simpering faces or cheerful voices will greet the thousandth customer as gracefully as the first. They will not take wasteful bathroom breaks, and they will grin through the pain of their exertions. Though they are loyal to their squadron, they will pretend not to notice when their comrades go missing in action; they will know that they themselves can be eliminated by friendly fire, easily replaced by someone else according to procedure, and they will accept this. They are interchangeable units. They will perform emotional labour, as in the words of one call-centre worker, "smiling down the phone" at customers.[19] Their role is not unlike that of the good wife or mother who sacrifices herself willingly and performs a labour of love for the benefit of her family. The relations of service work in a global economy not only conceal the labour itself (for ideally it appears effortless) but also make its fundamentally gendered, raced and classed dimension invisible. The skills of female service workers are not recognized as any more than natural talents; the service work of women of colour is seen as part of the natural organization of society.[20] Certain groups of people, marked by structural signifiers do not just earn a living by providing services to others; they are, foundationally, viewed as service providers whether or not they are paid. Whether they are mothering in the private or the public sector, it's all the same.

*

Service work is expanding, and now includes IT work. Indeed, most of the work associated traditionally with IT is service-based. This focus on technological services obscures the "life cycle" of IT that begins with low-paid, often female-dominated work in the manufacturing of IT components such as circuit boards, often (but not always) in economically disadvantaged regions. Indeed, in the 1990s, the thriving IT services industry of Silicon Valley depended on the "nimble fingers" (and, more importantly, the cheap labour) of Hispanic female factory workers. Often these women sweated over technical piecework and industrial solvents in factories not far from multimillion-dollar IT enterprises. Such invisibility, both of the workers and the objects they produce, adds to the illusion that IT creates wealth from nowhere.

Canadian data indicate that service work in the IT field is growing rapidly, with much of the work concentrated in Ontario.[21] The number of jobs in IT industries increased by 43.1 percent in the period from 1995 to 2000. This is over three times the job growth of the economy in general. In 2000, 3.9 percent of all Canadian workers were employed by the IT sector. Most of these employment gains have occurred in software and computer services, where employment is more than two times higher in 2000 than in 1995.[22] Given that the IT field is associated with a high level of skill, and the service sector a low level of skill, and that both of these associations are built on relations of gender, race, class and other structural markers, it will be interesting to see how certain types of IT work become redefined.

For one thing, service work is likely to become the normative standard of work, and the precariousness, repetitiveness and lower status that tend to characterize service work may accompany this standardization. South of the border, an MIT report notes that "the ways of the technology sector, far from being atypical, may actually serve as the model the American economy follows during the 21st century."[23] The report identifies a few reasons for this: the proliferation of high-tech districts, the growing importance (for whom?) of innovation, rising technical requirements of previously lower-skilled jobs such as manufacturing work and the demographics of an aging society.[24] However, the report also notes that despite advantages such as interest, creativity and positive challenge, service work has pronounced disadvantages, particularly a lack of job stability.[25] Companies have experimented with various models of service work designed to optimize flexibility, which has proven positive for workers who can enjoy working at home, use their non-standard hours to meet other responsibilities like domestic work and perform work that requires ongoing learning. However, this flexibility has come at the expense of job stability and security, as well as at the worker's diminished recourse if she is experiencing employer abuse.

Clearly, we are at a crossroads in terms of IT 's promise for a better future. One path leads to wondrous possibilities for new types of jobs that are exciting, innovative and well compensated. While there have been large-scale layoffs, there has also been consistent employment growth, and according to most sources there continues to be jobs available for those with the required skills. Although it has become more moderate, the pay remains above average. Workers in the Canadian software and computer services industry earned on average $52,565 in 2001.[26] In comparison, in 2001, the average salary of a full-time male worker across all industries was $49,250, while that of a full-time female worker was $35,258.[27] Some data suggest that professional work in technology is the fastest-growing sector (while lower-end jobs are swiftly being eliminated).[28] These factors have significant implications for groups of workers that have been traditionally marginalized because of social inequalities.

And yet there is another path, which has already been well-trodden and leads to the redefinition and revaluation of IT service work under global capital so that no matter what actual tasks are being performed, social relations determine who is performing them and how that is valued. This path leads to the "feminization" of particular sectors, rendering certain types of work low-status and low-paid, regardless of what the work is. It leads to the skill expectations for this work increasing while the pay and status do not. Most importantly, it replicates and reifies the divisions and stratifications by structural relations that have historically shaped the labour market and work experiences of female workers. Which path will female IT workers take? To interrupt the metaphor somewhat and provide a possibility that defies the laws of physics, I believe that they are quite likely to take both paths in varying degrees. Tari, for example, manages to cobble together a mixture of satisfying, creative and self-directed work in the form of self-employment combined with more menial lower-paid jobs. While nurturing her own small Web development business, Diane works part-time as a mystery shopper to supplement her income; however, because she is considered to be an independent contractor in that job, she cannot enjoy the regulation and benefits of a full-time employee. We may see some groups of women trot down the first IT job path, enjoying the scenic journey, while other groups of women, disadvantaged by structural factors, will have no choice but to trudge along the second route, which never again converges with the first.

*

Call-centre work is implicated in the gendered history of technological development. First, there was the telephone, initially intended as a tool for businessmen but quickly colonized by women wishing to communicate with others, as well as to find gainful employment as the soothing voice of connectivity. Then there was the typewriter, whose design had to be modified to slow down the swift fingers of female typists in order to stop them from jamming the keys. Finally, the computer united these two technologies in a metaphoric sense, enabling new ways of working and doing business, and this union is fundamental to understanding call-centre service work. Mated with telephones, computers can match calls to callers, ensuring that the caller always deals with the same agent, and that the caller's file is displayed to the agent immediately. Calls can be routed, preferentially distributed and tracked by volume, duration and "abandonment rates" (how long a customer will stay on hold before disconnecting). This implementation of technological prowess is not accidental; rather, "the driving force behind the decision to establish call centres ... has been the pursuit of competitive advantage."[29]

As with telephone operators and typists, the typical call-centre worker is more likely to be female and working part-time, since part-time work in this field is used as an employer strategy both to respond to peak demand times and to reduce the likelihood of worker cohesion and security.[30] Femaleness is also seen as an advantage where the worker is performing "care work" and "emotional labour" over the phone, soothing ruffled customers and seducing clients into purchasing goods. Call-centre workers' job advancement and salary are limited, as the relatively flat organizational structure of the centre precludes significant promotion. Workers are transient and burn out quickly from the stress and working conditions. "Encouraged attrition" in government-sponsored call centres helps employers to artificially inflate their job creation numbers. Call centres deliberately exploit a difficult job market, squatting in areas with high unemployment. In a display of co-dependence, depressed economic areas advertise their wares. Come to Kingston, come to Hamilton, come to Sudbury, come to North Bay. Our people need work, mayors plead. It seems little different from the "company towns" that sprung up around mines, fisheries and single-industry factories not long ago, though this time towns are selling their attractiveness, communication ability, soft skills and pleasure in serving, rather than their ability to provide craft and goods-producing labour. It is a feminized vision of labour's future.

IT INSTRUCTION: THE ROUTE TO GOOD JOBS?

In 2001, a press release from the Association of Canadian Community Colleges trumpeted the leadership role that educational institutions would play in shaping the skills of the IT workforce. The document is full of preening can-do spirit:

> Canadian colleges and institutes are innovators and leaders in bringing highly qualified professionals to the workforce. Internet products, solutions, and the full spectrum of e-potential can be sourced through the state-of-the-art education provided by post-secondary colleges and institutes from across the country.[31]

In 2003 I scan a want ad, which asks for at least two years of a computer science or computer engineering degree and in-depth knowledge and experience of database development, networks and several programming languages. The wage offered is $12 an hour.

Perhaps in order to make their sales pitch about the appeal that that their certification will have on the job market, many educational institutions provide data on how many of their students find jobs in their field and what they are paid. George Brown College, for example, in a spring 2004 poster advertising campaign, boasts that 91 percent of its graduates find employment. Its Web-site copy, however, is somewhat more vague, stating that "job-ready graduates achieve a high degree of success meeting career goals after education." I crunch the numbers provided by Niagara College, which offers various certifications such as computer network operations and computer programmer-analyst. The outlook is lukewarm. On average, a 2003 graduate from one of these programs earned about $30,000, less than the national average for both males and females in full-time, full-year jobs. The default rate on student loans for these programs in 2001 was as much as one-quarter of students. While 2003 graduates managed to find jobs, it was not always in their chosen field. The most successful were computer programmer-analysts, 80 percent of whom were working in their field, but only a surprising 56 percent of computer programmers were. Gloomier still, these numbers are worse than two years earlier. The average salary is $1,000 lower and the work placement rate for computer programmer-analysts is down by 5 percent. One Toronto university has just implemented a mentorship program to assist its graduates in job hunting. As one mentor notes, "Four or five years ago, most of the graduates in IT would have gotten jobs straight away — now there's a high non-placement rate."[32]

I speak to a kind but somewhat harried student placement officer who is also an instructor at another one of the IT colleges. I ask her how students fare in the job market once they have completed their certifications. She sighs. "Well, we could *definitely* do better. Placements are always a challenge." With the current economic climate, graduates are having great difficulty finding jobs. The jobs that they are finding are at lower salary levels (she guesses $33,000 to $38,000), at lower status levels (such as helpdesk work) and are of poorer quality. "There's a *huge* difference in the incidence of contract work as opposed to full time. Students are finding related work, but on a contract and a part-time basis." Is this an issue of the quality of the education provided? She thinks not. "We have a lot of bright students. We have a good name in the IT field. Our staff are current, constantly reviewing and upgrading their skills, and we respond quickly to the market demands of employers. The feedback we're getting from employers is that certifications are *extremely* important." She has noted that the face of the student body has changed. Now she is seeing older students with established families, often with university degrees, often coming from other work experiences. She esimates that perhaps one-third of enrollments at the college are post-graduate students. Despite the demographic shift in terms of age and work status, she still sees few women. "We're very low in the female population, but we're working on that."

Tari, who both completed certifications from and teaches at IT schools, is characteristically cynical. One problem she notes with IT instruction is that the instructors are also in precarious employment situations. She feels that the practice of just-in-time, flexible manufacturing has been extended to curriculum development, with instructors asked to respond rapidly to changing market demands and develop courses quickly. Remuneration can be poor, although it varies from school to school. Hours can be sporadic, scant and often in the evening. Instructors are often promoted as "working professionals," which is seen as advantageous because they should be experienced in their field. In practice, what it can also mean is that schools who hire them pay them hourly, with no benefits. In one school, she says, instructors have almost completely been eliminated and students pay tuition to do self-study using books and CD-ROMs.

However, it is for private IT schools that Tari saves her greatest ire, believing that they fabricate evidence of the need for IT workers. "As far as I am concerned," she says, "this whole thing about there being a shortage of IT workers is a crock." She points to many skilled IT workers, people like herself, who are under- or unemployed, as evidence to support her argument. She is concerned that many small private IT colleges are unregulated by the Ministry of Education and are not accountable to their students in offering short, application-focused programs. Her words are blunt. "Most of the programs were, and most of the programs still are, worthless." She feels that the focus on pushing credentialed IT graduates into the workforce quickly means that they don't develop transferable skills. "People continue to pay big money to go and learn how to use software and/or programming languages, but not how to really design or really program, and worse, they don't learn how to problem-solve, something designers and programmers must be able to do, and something which can't be taught while cramming in multiple software programs and programming languages over a short period of time." She warms to her subject, and wastes no pleasantries. "These schools, private and public, are lying in their ads. Lying through their teeth. Their admissions people are basically sales people ... They can smell fear and desperation like sharks smelling blood in the water. If they weren't selling hopes and dreams to the unsuspecting, they'd be pushing swampland in Florida or farmland in Afghanistan." In a climate where education is a commodity, people need to be critical consumers. "People do more research into buying an appliance than enrolling in educational institutions."

Laurie, whom we met in chapter 2, has been in the IT industry for several years. She notices a trend towards privileging formal IT certifications over actual skills and experience. She finds that employers are now making unreasonable demands on employees for certifications, without looking at practical expertise. "Companies have become very picky, in terms of what roles they hire you for, what skills they want you to have. They want the earth and the moon and the stars. I had hoped to have an interview with a company, but basically they refused to interview me because I didn't have a degree, even though I have an honours tech certificate and ten years of practical experience. And the money they're offering has gone down too." She is now enrolled in an IT degree at a Toronto university, in order to improve her chances. When I ask her about what changes she has noticed in the last ten years, she identifies the importance of specialization and credentialization. People used to be generalists, she says. Now, "with all those certifications that you can get, people are asking for them. Also, there's all these specializations that never used to be there."

While she's enjoying her studies, she's a bit disappointed that companies didn't seem to care about her abilities. "I've noticed over the years that I'd be passed up to do a specific role in a company even though I had the skills, because I didn't have the piece of paper. The person who had the paper but not the skills was given the job, but couldn't do the job. There's this pervasive idea that this paper gives you some kind of magic knowledge that really isn't there. The theory is good to have but you need the practical application." Laurie is somewhat equivocal about the role that age plays in her success. She is pleased with her ability to focus and apply what she has learned, which, she feels is a result of maturity and years of hands-on experience. But Laurie also worries that her age works against her in an industry that privileges youth. However, she also points out that the speed of IT instruction is a liability. She is concerned that her IT degree is being promoted as a quick route to a senior-level job, and that students don't realize the importance of working in the field over a longer period. "It takes a lot of time to learn. With this program, it's such a brief overview, and they say you're qualified to run a department when you're done. I say yeah, maybe after ten years. Start on the ground floor and put in your time, get some experience under your belt, otherwise there's no way. I couldn't do as well as I am doing here at school if I hadn't had the practical experience."

Experience, intuition and problem solving are key, says Laurie, and these things are built, not bought. "In IT you have to be able to think on your feet, because the solutions are not obvious. Quite often the answers are found by accident. You don't know why it works but it does." Speed negates quality, haste makes waste. Even in the faster, better technological age, good quality learning still takes time and patience.

DIVISIONS AMONG IT WORKERS

Industrial capital has always depended on the creation of divisions among workers, and IT is no different. This is a productive strategy, for it inhibits collective action and ensures that mutual resentment among groups of labourers will divert attention from the consolidation of employers' power. It also obscures the fact that high-level IT workers depend on the labour of low-level IT workers. For example, devising and implementing the architecture of a computer network depends on the presence of a connective infrastructure. Someone has to crawl into a small, dirty space and install fiber optics. Someone needs to solder the electronic pieces of the circuit boards together so that the computer works. Someone even needs to pick the coffee beans, or work behind the Starbucks counter, to fuel the late-night programming. These are all valuable tasks, though socially and economically they are not all equally privileged. The stratification of the workforce has traditionally occurred along axes of gender, race/ethnicity and socio-economic class, and it is bottom-heavy. A thick foundation of menial workers is required to support the sparser top layers.

IT work used to be grunt work. The first "computers" were women who calculated ballistics tables during the Second World War or who used "comptometers" — mechanical calculators — as part of accounting functions.[33] Because this work was considered tedious and low-skilled and was associated with clerical functions, it was allocated primarily to women. With the emergence of computer science as an offshoot of other formalized science programs in academia, IT skills and work gained a new value. Still, computer users laboured in obscurity and under a cloud of social disdain until the emergence of the PC suggested new economic opportunities for lucky geeks. In the 1990s, certain types of skills, because of their scarcity as well as their possession by certain groups of people, were considered important.

Nearly anyone with a shred of computer aptitude could hang out a shingle as an IT "envisioneer." Currently, relations of IT work and skills are undergoing another fundamental transformation. IT work has been incorporated into networks of globalized, transnational capital and is becoming more diffuse and distributed. Consequently, it is also becoming increasingly divided between a "knowledge class" and "technicians." The former are professional IT workers, situated at high levels and often working within scientific, engineering and academic fields. The latter are the new, emerging IT proletariat, commanding lower status, lower wages and greater job instability. While they demonstrate varying degrees of skill in their craft, it is primarily their fluid labour power that is attractive to employers. They represent commodities that may be purchased and sold at whim, and to which the employer owes little responsibility. It is the former group that we tend to imagine when we think of the IT field. It is the latter group that may compose the bulk of the reality. In the next chapter, I look at this theme of continuity through change: ways in which work has changed while remaining the same.

CHAPTER 4

New Work versus Same Old, Same Old

SOME YEARS AGO, as a starving undergraduate student, I was desperate for employment. I responded to a job ad that promised good wages, on-the-job training and flexible hours. Wearing my nattiest suit, I showed up for the interview, held at a low-rent office building on the outskirts of the city. I was herded into a fluorescent-lit room full of other people sitting on folding chairs, all looking rather nervous and dressed in their best. Once assembled, we were told by an efficient-looking execu-droid that our job would involve selling long-distance and cellular phone services door to door. We would be trained, but not paid for training. We would work solely on commission, receiving a couple of measly dollars for every few subscriptions acquired. We would be required to wear company jackets, but have to buy them ourselves.

I began eyeing the exit.

However, I had no chance to make my escape, as the person who clutched my carefully composed résumé quickly shuffled me into an individual interview. In this interview, the conditions of the job were reiterated, but glossed over with promises of the gobbets of cash I would make. According to the recruiter, I would be rolling in piles of money by the end of the month. As soon as the interview was done, I ran and never looked back. I did not show up for unpaid training. I did not buy my red company jacket. I did not make a fistful of dollars, but somehow I suspect that wouldn't have happened anyway.

That company subcontracting sales work in early 1994 was Sprint. In 2003, the telecommunications company under fire for the same practices was Rogers. I was to sell long-distance and cellular subscriptions; Rogers subcontracted workers to sell high-speed Internet and cable services.[1] Both companies share a common practice: the subcontracting of menial sales work for low pay and in poor working conditions, making invisible the cheap labour of promoting the IT industry. More importantly, both companies depended on the arm's-length relationship of subcontractors to disavow responsibility for this.

In the case of Rogers, many subcontracted workers did not receive any wages or received only a tiny portion of what they were owed. Though subcontracted, these workers had been trained at a Rogers location, by a Rogers staff rep. When on the job, they wore a Rogers badge identification. Yet Rogers claimed no responsibility for the actions of its subcontractors. Not surprisingly, these subcontracted workers are also generally new immigrants or people desperate for employment, people whose precarious employment situation and social location rendered them vulnerable. I was lucky enough to have the option to walk out on Sprint. Many other people would not. A few decades ago, these people might have been garment or industrial homeworkers instead. Subcontracted IT work repeats the same cycle of uncertainty, poor pay and working conditions, and worker distress. The content has changed, but the form has not.

*

One of the most pervasive discourses about IT work, as I have mentioned in previous chapters, is that it represents an unprecedented, new form of work. In some ways this is indeed true. Technological facilitation of particular tasks has enabled people to do work in different ways. Meetings can be conducted by teleconferencing, IP telephony or on-line chat. Documents can be shuttled from place to place instantaneously, disregarding the constraints of space. It is possible for the time and space of work to be reconsidered in positive, productive ways. Asynchronous work can enable people to work on their own schedules, focusing on task performance rather than obligatory "face time" at the workplace. It can also mean that workers from various locations can participate in collaborative projects, attending meetings virtually and exchanging material electronically. This has the potential for new kinds of work practices and new kinds of work alliances to be produced. It can challenge hierarchies of physical space and provide employment opportunities for people living outside of busy IT hubs like Toronto. For example, Janice, a process documentation specialist, is able to enjoy living on her farm in southwestern Ontario while still working at a technical company. According to her, this is an excellent arrangement, as she says, "I can work for a high-tech company while living out in what most people would consider the 'boonies.' I can be with friends and family, and have a challenging job at the same time. It doesn't get much better."

However, in other ways, IT work can replicate old ways of doing business. The restructuring of work, which has a technological component although technology itself is not primarily responsible, has resulted in significant shifts in work practices but often with a minimal increase in the power of the worker to control her work. The home-based IT work of today has suspicious similarities to the home-based "sweating" work of a hundred years ago. IT corporations remain resistant to worker advocacy and activism.[2] Like the scabs of yesteryear — themselves disadvantaged workers who needed the money — who were brought in by employers to challenge the unions, outsourced workers in lower-income geographic regions are employed to cut labour costs without regard for equitable compensation and working conditions. The "virtual organization" often looks very similar to the standard corporation; merely the bricks and mortar have vanished.

Structural relations of power and privilege mediate how the time and space of IT work are experienced. Human life-time and life-space are increasingly abstracted and commodified.[3] This in itself is not new. Control over the time and space of work has been an ongoing struggle between workers and employers for centuries. What is new is the scale on which this contradictory space-time can occur. The space of IT work can be a non-space or an everything-space of cyberspace. It can be as small as the circuits on a microchip, as familiar as home, or it can span the globe. IT can provide asynchronous time, with work performed in disparate schedules, or it can create new kinds of synchronous time, connecting people instantaneously to their jobs and to one another.

The space of IT work is not experienced in the same way by all workers. For a low-income teleworker, the home may represent a confined space where time is regulated by the technological monitoring of the employer, and where demands on time are made by both paid and unpaid labour responsibilities. Flexibility to this worker may simply represent more hours of labour, merely at different times. It may also mean flexibility for the employer, to hire and fire, to reassign duties at will, to be free from the obligation to provide any benefits, to lay off workers when business is slow and have them work overtime when business is up. For a high-income professional, the home may represent a pleasant, comforting space where work may be done at a time which is chosen and convenient, or at least determined to some degree by the worker. Flexibility to this worker may represent the ability to do work when and how she desires, and the freeing up of time to pursue other enjoyable activities. The time and space of home-based labour is experienced in a gender-specific way: for women, for example, who continue to perform an inordinate amount of unpaid domestic labour and who experience gendered expectations about their role in the home, homework may not be a fulfilling, exciting job but merely another form of gendered drudgery.

Technologically facilitated work has both expanded and contracted work space and time. For many IT workers, work time has increased in its volume, intensity and density, but it remains confined by the limits of human ability — after all, there are only so many hours in a day. In high-school chemistry we were taught that as the temperature of a substance rises, its volume increases; if it is trapped within a container whose boundaries do not change, the pressure also increases. IT workers required to do more work with the same amount of time often find themselves crushed up against the margins of this finite capacity. For other IT workers, time gapes open as they struggle to fill it with their reduced or downsized work. They may cobble together part-time, freelance, contract and temporary work in order to create enough worktime to sustain themselves. Time and space form a series of relationships. In the case of IT workers, certain types of working time are facilitated by other types of working time; work done in certain spaces is enabled by work done in other spaces. Relationships of temporal and spatial location intersect dynamically with relations of social location.

The rhetoric of IT work, as does the discourse of post-Fordist economic apologists, obscures the fact that while the actual incarnation of this work is facilitated by technology, these structural shifts are actually part of larger trends in labour markets and economies.[4] To quote Yogi Berra, it's déja vu all over again.

Standard and Non-standard Work

Standard and non-standard work differ in time and space. Standard work is performed in a formally organized space, using some kind of insitutionalized schedule, while non-standard work is often performed in "marginal" spaces: homes (the worker's own or other people's), fly-by-night businesses, other workplaces, spaces where the workforce is transient and unstable, spaces that can disappear unexpectedly. Non-standard time is chaotic: split shifts, irregular shifts, part-time, on demand, initiated at a moment's notice, terminated without warning. Many of the women introduced in chapter 2 spent their time this way. Often they got up early to do a couple of hours of work before their children woke up or stayed up late after the children went to bed. They might have done a few hours after their "real" job, to make a few extra dollars from contract jobs. Karen, for example, works full-time during the week, then works on contract doing Web design on Friday and Saturday nights. "My boyfriend works the night shift then," she says, "so I spend the time doing the work. I work from about nine in the evening till four in the morning."

So-called non-standard employment has historically been and continues to be so common that the term "non-standard" is a bit of a misnomer. While the term is often applied to cover a variety of practices, it generally refers to some kind of deviation from a putative norm or "standard" of hours or type of pay, or both. "Standard" employment is often thought to be something along the lines of what we refer to as "9-to-5" work (though it may actually be done in shifts), which is undertaken permanently, on a full-time basis and year-round, with consistent salaried wages given in a "public" format (i.e., declared on a company payroll, not paid in cash under the table or a similar arrangement). This normative model is implicitly founded on the assumption that the work is performed by a white male tradesperson whose work is organized and standardized.[5] The standard employment relationship, then, is less an expression of actual work practice and more an expression of idealized or normative work practice. It is somewhat misleading to conflate all forms of non-standard employment, because they vary in terms of where the work is performed, the consistency of the wages, the ownership and distribution of the means of production, the level of control over the pace and quality of work, and so forth. There are many forms of non-standard work that are preferentially performed by women, primarily part-time work, temp work and homework. It is important to address non-standard work because it is becoming the norm of IT work. Some might even say that IT work is, by definition, non-standard work. Certainly for some workers, the exuberance of the early dot-com years was based on the rejection of the standard work arrangements of the previous generation.

However, non-standard work is acquiring a more sober character in the face of globalized restructuring. Restructuring is part of a worldwide shift in the organization of work which involves "the collapse of work itself," underemployment, jobless growth (increased output without necessitating worker-producers), the deinstitutionalization of workers and work, computer-enabled worker surveillance and deskilling paradoxically combined with credentialism (the shift from informal networks of knowledge to formally qualified skills). Technologically based restructuring results in two central developments: the substitution of machine and automated control over people for human management and administration, and the self-reflexive nature of computer systems that are able to self-regulate independently of the intervention of human operators.[6] This organization of work is not neutral (as we have seen, technological intervention is never neutral) but has gendered consequences. With technologically facilitated restructuring there is "a polarization between computer-assisted and computer-controlled work along clearly gendered lines,"[7] with male-dominated data-processing jobs enjoying greater autonomy and status than female-dominated data-entry jobs. Professional-managerial workers enjoy the analytic and strategic power accorded to them by technology, while lower-status workers such as clerical workers submit to the demands of the machinery.[8] For women, this type of technological work reorganization generally means a move towards permanent non-standard employment, particularly home-based work in traditionally female-dominated areas of clerical, sales and service work. Surveillance is a significant component of this type of work, with technologically based process assessment (often known as total quality management, or TQM) considered the most advanced form of management.

Today, the North American economy is moving towards an increased reliance on non-standard labour. There has been a decline in manufacturing jobs and an increase in jobs that are service-based, as we saw in the previous chapter.[9] Far from being accidental, non-standard employment is a strategy used in a particular way in a particular economic context. Non-standard labour, as with all other forms of labour, is structured by social location. Not only does the feminization of labour parallel the rise of restructuring and non-standard employment, but it fundamentally constructs the relations upon which economic shifts are based. Put another way, global restructuring is intrinsically founded on the feminization of labour.[10] Non-standard employment in many forms is not only performed by women preferentially, but shares numerous features which distinguish it as "female work": the casualization of employment, the low status and low pay and segregation in which gender intersects with social location to result in income polarization.

FLEXIBILITY

A common theme in current conversations about work is flexibility. It is a messy term, used to denote a variety of things: the casualization of the workforce, the capacity of companies to adjust to shifts in the marketplace, the ability of workers to learn diverse tasks and to respond to changing employer demands. In the case of IT work, flexibility is often seen by employers and workers as a defining feature. It is assumed in pro-flexibility literature that the technology itself will create the conditions whereby workers must continually adapt to rapidly shifting skill requirements, while the work will be challenging and novel — a refreshing sorbet to the tired palate of the worker mired in the standard employment relationship. The worker can and will work anywhere, any time. This is inescapable, obligatory and progressive.[11] Oddly, the temporal elements of this fantasy of new technological work echo pre-industrial forms of work that were based on the irregular cycles of agricultural or craft production and that irked rationalist factory owners in the eighteenth and nineteenth centuries. Indeed, the early ideal of capitalist production was anything but flexible: work was increasingly fragmented and monitored, partitioned and procedurized according to quasi-mathematic formulae of "scientific management." Notably, it was also technology that was seen to enable these conditions. Workers had to emulate the machines they worked with, and the rhythms of steam engines and assembly lines were precise and unforgiving. Today, of course, things are different. Individually (not collectively) "empowered" workers are casting off the shackles of mechanistic mass production in favour of liberated, creative agility.

Or are they? The ideal of rigidly controlled production, whether it ever made it to practice or not, was nevertheless applied only to particular groups of workers. While women were sometimes included in this group, as they formed the bulk of the factory workforce in some areas, they were often excluded from the union protections accorded to male trades or craft labour. The majority of female workers, immigrants, younger workers, homeworkers, casual labourers and other marginalized worker groups have always been acquainted with so-called flexibility, only they might have called it precariousness, underemployment, lack of worker protection or "making do with less." The revelation of flexibility has less to do with technologies and much more to do with shifts in employment patterns and discourses. "It is a conception," notes Anna Pollert, "legitimising pliability, insecurity, unemployment, and 'getting by' with self-employment."[12] It is also notable for its blindness to social relations of gender, race/ethnicity, age, class and geography.

*

Rebecca is in her mid-thirties. She has worked at various jobs in a number of different types of companies: in non-profit startups, at multinational IT corporations, for management consultants, at an electronic art gallery. During her art history and museum undergraduate studies, she became interested in how software was being used in museums. Before the introduction of the Web, she was designing and using databases for the management of museum collections. She recalls, "We used databases for collections management, to keep info about collections, HyperCard programs, visitor interpretation programs. Some museums were just starting to set up computer kiosks, where they could get more information, experimenting with interactive features." This piqued her interest in software design and development, particularly in user interfaces. "I've always been interested in how technology can refine work processes, make them easier and more efficient. I found SAP quite interesting, like data mining, if I ever went into that area, it would be through an MBA with a focus on marketing informatics."

When the Web came along, Rebecca was very excited by the possibilities. "I'm not hugely interested in the whole hardware aspect, other than theoretically and generally. I like to know what's coming down the pipe, but you ask me to read a manual on routers, I'm not interested." The Web enabled her to build on her interest in representing and conveying visual information to users. She says enthusiastically, "I love the vividness of colours on the computer screen, I love the shifting and transparency of Flash. I like being in a milieu where people are thinking about how to enable communication and information management."

Eventually she was able to translate her interest into working for a startup. She found that her mix of youth and maturity appealed to her employer. "Because it was a startup, they needed someone dynamic and young thinking and flexible, but they also needed someone who could talk to folks who needed stability, like venture capitalists or government. They couldn't have had a twenty-two-year old guy with a goatee going in and talking to the minister. They needed someone to bridge the startup and corporate world." She sighs. "That lasted till our funding angels ran out of money." By the time the company got going, the dot-com bubble had burst. Investors were cautious, and nobody had any money to share.

Currently, Rebecca is teaching ESL because of the difficulty in finding work. Part of the problem in finding work now, she says, is that it is so much easier to outsource to other regions, or hire young people who don't mind long hours for relatively lower pay. "The programmer in Bangalore can make six thousand dollars and live very well. Also, the type of person who really loves coding and will sit there till two a.m.

and order in pizza and drink pop is someone who's twenty-two and doesn't have much else in life and loves to do it. Someone who's thirty-five or forty, has a family and mortgage, hopefully they will have found more in life than lines of code and won't be so eager to make sacrifices for the company."

The theme of flexibility comes up repeatedly. I ask Rebecca what her actual experience of flexibility has been. While there was a danger that flexibility could result in a lack of definition and goals, she liked the flexibility she found in technical workplaces. On the one hand, people were open to new ideas and often not as invested in corporate hierarchies. In new media companies, workers tended to be younger, more urban, hipper. On the other hand, she found that flexibility often meant staying till midnight, often working unpaid hours, in order to be a "team player." However, she says, "I rarely turned down work because I knew if I didn't do it someone else would." The speed at which work got done often reflected the late adolescent jubilance of the dot-com spirit, as well as the mandate of "just-in-time" production. "The speed of the industry, before the dot-com crash, the days when *Fast Company* magazine was three hundred pages thick, people even talked fast. The speed of the industry really was excessive. I thought that either these people are thinking faster or they're just so hooked on looking fast, they're not considering the ramifications of their plans." Often, she said, speed was fetishized over sensibility. People would make quick plans because of the pressure for speed, or because they wanted the exhilaration of working at breakneck pace, but their plans would disintegrate in the face of practical problems. Rebecca preferred to take her time and think things out, a characteristic at odds with the rushed climate. "There's only so fast that you can work. Sometimes you have to sit down and think about things. So many times we had to fix things, or things were done sloppily, because people didn't do them well and they had to be reworked." The just-in-time model, which emphasizes immediate execution of all phases of the production process and delivery of needed materials moments before they are required, is ostensibly dedicated to eliminating waste. In fact, suggests Rebecca, it is still haste that creates waste.

Despite the outward emphasis on "flexibility," employers remained focused on the bottom line. Flexibility in many cases meant creative, if not fraudulent, accounting practices. According to Rebecca, it was common to oversell products in order to be able to record sales, much to the horror of the technical personnel. "The developers would say, 'We can't do that,' then the marketers would say, 'We sold it to the client, we promised it three months down the road, and you have to do it.' Then the developers would be on the hook to develop something. The managers of the company knew this was going on, but they tried to keep everyone humming along because they had to keep showing profits every quarter."

An emphasis on speed above quality, says Rebecca, is short-term thinking. Humans aren't machines that can be endlessly upskilled and upscaled. The notion of limitless exponential growth is like expecting a teenager to continue their growth spurt until age ninety. This approach, she tells me, causes people to burn out and companies to be unfocused. "When you look at nature and human cycles, there are ups and down and highs and lows. You can't keep constantly going higher and higher and higher. You have to step back and focus and breathe. Nobody was willing to take a breath."

HOMEWORK, TELEWORK AND THE "VIRTUAL ORGANIZATION"

Patricia is a single mom in her fifties. She has a grade nine education, and currently runs her own business as an IT project developer. Though she doesn't have formal credentials, she is a fast learner and was able to pass the entrance exams for a general science course at the undergraduate level. However, she found school too emotionally stressful and dropped out. After a series of jobs, which included zoo keeper and a denturist practice, she decided that self-employment was the right path for her. She was inspired to get into IT because of her children, who have learning disabilities, and purchased a computer so that they could type rather than write. The presence of the computer in the household made her decide that it was worth attempting to use. In keeping with her independent, freelancing spirit, she dove into self-instruction. "So there was the computer, and I said I'm going to learn to use that thing. And basically that was it. I crashed it about fifty or sixty times, and opened and closed every single program, and took out the modem, and put it back, and did everything I could until I knew how it worked. That's the kind of person I am. If I learn something, I want to know everything about it, right from the bottom up." This was particularly challenging in the early days of computers, she stresses, because operating systems were not "plug in and play." One needed a good deal of know-how in order to get a computer working properly. "Back in those days when you went on the Net, you were given a phone number, and you had to phone in and download programs, and totally configure your modem and everything else, all by yourself. Nothing was automatic. I started with bulletin boards, there was no real Internet, so it was just out of sheer determination that I wanted to learn how it worked." She considers this perspective to be characteristic of freelancers who, she feels, "have to be extremely flexible, adaptable, roll with the punches, and have a lot of initiative to learn new things when they come up."

Learning to use her son's computer paved the way for Patricia's present occupation. She is thrilled by the potential that her current work offers for global interaction. IT has given her the ability to gain clients from all over the world. Her virtual space has expanded to include the global village. "I get all my clients on the Internet, strictly. I work virtually, I don't work locally at all, not because I don't want to, just because there's more work on the Net." Patricia is able to bid on projects located anywhere. She was a relatively early adopter of the telecommuting concept, and as a result has built a solid foundation of clients. "I was ready to telecommute several years ago ... so I have clients in Germany, I work with people in Great Britain, I have clients now in Ireland, in the United States, all over the place, and it's all done electronically." For Patricia, working remotely is a wonderful experience, both personally and professionally. "You meet people all over the world, all these contacts and connections. You just learn so much, so to me that's something I enjoy, it's a personal benefit. You build up a network, in a completely different way, that's much broader and gives you a good feel for what the world is made up of. I like that rather than just doing local work or working in Toronto, it's exciting to know you're doing business with someone in Germany or Ireland, and to interact with people on that level." She is excited about seeing what the technology has to offer her in the future, now that its use has become normalized. "The technology is the means by which things are accomplished; it's not the end result. When people become competent with technology, it's just like using the telephone, you don't even think about it anymore, you don't talk about using the telephone or dialling, that sort of thing, but we have to go back to, okay, now we understand what it is we're using, how are we going to best use it? Now that we're all savvy, what's next?"

Like many self-employed IT workers, Patricia works primarily from home. Though she greatly enjoys this and has her own office, she admits that there were difficulties establishing her space in the beginning. When she first started, she says, her kids would say, "'Oh Mom, you're home, you can do this or that.' I just laid it out flat right at the beginning: I'm home, this is important to me, it's going to succeed, so make your own." She laughs. "They're old enough now. The business is my income. It doesn't come first, my children come first, but they have to know that there's a definite boundary there, and they're old enough to understand that if I don't work and have an income, they'll have nothing to cook." Patricia feels good about her work because she has clearly defined spatial boundaries in one sphere but has transcended them in another. She has carved out her workspace at home, but nevertheless enabled herself to work virtually in a positive way.

Though Patricia clearly identifies advantages in the dissolution of national boundaries in terms of virtual work, she does, however, worry about how work will shift. She already spends a great deal of her time looking for work and maintaining existing contacts. "It's like looking for work constantly," she says, "beyond your normal job hunts, because many of your contacts are small. You probably spend at least thirty percent of your time hunting. At least I do." Now, with the emergence of telework in other locales, it may become more challenging for her to survive. "The Internet is changing freelancing, because it does open up the opportunities globally. There's a lot of competition and I see big changes coming in the very near future, because without question, the Eastern countries are going to be offering some pretty stiff competition to North America. I think anybody that has their eyes open is going to see that they have to work smart in order to stay ahead of that."

*

In conjunction with the trends in the economics and philosophies of doing business, advances in telecommunications technology have resulted in a shift in the physical spaces of certain kinds of work. As I just discussed, while women have always performed paid work in the home, it tended to be an extension of their domestic duties: taking in laundry, sewing, child care. Currently, the practice of homework has shifted to incorporate a variety of new practices such as remote data entry and technical support services. Like waged work done in the "public sphere," technologically facilitated homework or telework is diverse in form and content. Accurate assessment of telework is difficult because of varying definitions of the work itself. In some cases it refers to the work done by technical professionals who happen to be located physically at home.[13] In other cases it refers to work that is deliberately located in "private" homes as a means to expand the potential of sweatshop-type technical labour such as data entry. In the latter case, such a move is beneficial to employers as a means of reducing overhead and infrastructure costs, as well as a means for "increasing productivity" or extracting more work from the employee.

Telework's frequent location in the home, or at another site which is not the "official" workplace, can make it difficult to see and quantify. Teleworkers may be part-time, contract, full-time, freelance, self-employed or employed by a company, and work practices range from highly structured and formal to loose and informal. They may have a great deal of personal autonomy or have their work under close surveillance. Telework may be a convenient or mandated choice in relation to domestic work and child care.[14] We might also distinguish between telework and telecommuting. Heather Menzies uses the term telework to "describe people who are employed at home doing generally routine information-processing work, often under poor pay and working conditions, with little power to do anything about it." Telecommuting, however, refers to "people who have a regular office and occasionally work at home on a laptop or personal computer, usually under conditions of their own choosing." She argues that telework, which is "work downloaded by computer networks and completely defined and controlled by management information systems could become the new post-Fordist model of work, a pillar of a new cybernetics of labour."[15]

It is hard to determine criteria for teleworking versus homework alone. There are many kinds of home-based work, such as piecework, which are not dependent on information technology. However, there are also many kinds of home-based work which, while IT is used, would not be considered technology-specific per se. Technology has not been the primary factor in the development of telework in that only some forms of telework have been made possible by technological innovations and broad based dispersal of them. Various forms of both professional and clerical work were done at home long before the technological tools, such as e-mail, which facilitated these tasks. IT has thus become the facilitator, rather than the creator, of homework. Other factors, such as the interests of the employer, are what drive the actual work arrangements.[16]

Distinguishing between types of home-based teleworkers is not a purely semantic exercise, for there are fundamental differences between types of telework. In general, telework is polarized. One group of teleworkers, who tend not to be fully based at home but rather telework for only part of their working time, are likely to be male, professional and highly skilled. They are working at home because working at home is a privilege and a demonstration of their employer's trust and respect for their abilities. They may be freelance or contract employees, able to choose their work patterns with a fair degree of autonomy and flexibility.[17] A second group of teleworkers, who tend to be full-time homeworkers, are more likely to be doing low-status clerical work such as data entry or wordprocessing. They are also likely to be isolated from a larger workplace, and their employment resembles a subcontracting arrangement rather than the standard employment relationship.[18] They may even have been "downgraded" to telework as the employer sought to streamline their work force and eliminate job security, full-time workers and benefits. They are most likely to be female. In female-dominated white-collar sectors such as finance, regardless of how information technologies are implemented, they tend to eradicate the lowest-skilled clerical, managerial and professional jobs.[19] Low-end IT work like data entry is then substituted for many clerical functions, and this work is performed in "data entry campuses," often in economically depressed areas (for example, the Maritimes, and often in smaller cities or towns that are struggling economically), or in geographical areas worldwide with lower standards of worker protection.

There are also mobile teleworkers who may work transiently in assigned office spaces, or workers who are not explicitly teleworkers but who interact with their supervisors virtually (such as delivery people who are told where to go by a computer). The distinction between teleworkers is crucial, for lumping various types of workers together conceals the very real divisions between them. Telework is a polarized work practice, split between skilled male professionals and low-skilled female clerical workers. There is a vast divide along lines of pay, status, power, work practice, benefits, labour protections and autonomy.[20] By and large, the majority of the women I interviewed would be considered telecommuters rather than teleworkers, because of the relative autonomy and professionalism of their jobs. However, although many of them had ostensible autonomy and control over their work, in practice they found themselves highly dependent on the schedules and demands of clients and employers.

Advocates of professional telework or telecommuting argue that it provides numerous advantages for both workers and employers, although many of the benefits for low-status teleworkers have yet to materialize. The "virtual workplace" (which, in precise terms, rarely exists) has been lauded as a new structure that promotes creativity, flexibility and worker fulfillment and that eases the stress of commuting or interacting in the workplace. The need for costly office overhead is reduced, since the physical space of the workplace is diminished. In theory, virtual workplaces can increase worker networks ("virtual teams"), provide solutions for "outsourcing" or "contracting out functions that are not among the core competencies of the firm," and building a worker culture of co-operative and collaborative teamwork.[21] Others argue that virtual organizations "increase competitive capabilities," add "operational flexibility," have "improved communication and internal control" and have created "greater responsiveness to market (customers)," which includes "improved customer service."[22]

Many of the women worked from home at least part of the time. Some worked from home more or less full-time and were quite enthusiastic about the benefits of telework. Sandra, an IT consultant, loves having the opportunity to telework. Both she and her husband work primarily from home, and they find it an excellent arrangement. "We think it's improved our quality of life," she says. "We joke that we have a killer commute of ten seconds. We have arbitrarily set the end of our day as *Simpsons* time. Five o'clock comes, we break, and we go watch *The Simpsons* ... I like the fact that I can wear whatever I want; I like the fact that we have more money at the end of the month because we're eating at home, we're not eating in a food court. I get to set up my office with exactly the kind of equipment I want, I can play my music really loud, and do whatever I want to do." For Sandra, the flexibility of telework means increased choices and control over how she spends her time. "We like the flexibility that it gives us, that if there's something that we have to get done, some appointment that would normally be during traditional business hours, we just do it and then we make up the hours later in the evening." Autonomy is an important factor in this, and it is linked to job status. Sandra holds a very senior position. Her boss respects and trusts her enough to allow her to set her own hours and doesn't require her to put in what she considers unnecessary "face time" at the office. "The other thing I love about this," says Sandra, "is that when I'm between projects, I don't have to pretend to be working. Whereas when you're in an office, you still have to pretend like you're busy, and it affects everyone else around you. My boss knows that I'm going to do other stuff, but he also knows that when a project comes up with an insane deadline, I'll work till three a.m. to finish." For Sandra, the space of the home office is her haven.

Despite the fact that many people have positive experiences with telework, as a large-scale phenomenon, telework is generally more beneficial for the employer. The most common advantages cited by professional teleworkers are improved autonomy (for those that are self-employed and not dependent on a small oligarchy of clients), increased flexibility in working hours and greater convenience for meeting family obligations. Women, especially, may see home-based work as a solution to child-care dilemmas, whether or not this solution actually materializes (evidence suggests that it does not), while men are more likely to view home-based work in terms of the advantages of the work itself.[23] Absent from these plaudits for telework is consideration of the actual working conditions of the teleworker, especially the low-status teleworker (as opposed to the telecommuter or professional teleworker, or self-employed homeworker who facilitates her/his business interaction with technology). A large majority, nearly 86 percent, of Canadian teleworkers are not unionized. Nearly 60 percent earn less than $20,000 annually, and another 25 percent earn between $20,000 and $40,000 annually.[24] Few have substantial benefits such as retirement savings, medical benefits or dental plans. Even though many teleworkers might do the bulk of their work for one employer, they may still be considered as "self-employed" or as subcontractors and receive no job protection. The choice to use telework as an option to manage domestic obligations may shift attention and responsibility away from employers' demands on workers and from social programs that provide adequate child-care support. Thus, we must consistently ask "Advantages for whom?" when we hear of the many positive and ostensibly liberating attributes of telework.

These perceived advantages are not a natural outflow of the technology itself. Home-based work, especially for women, has a long history and has explicitly functioned to increase employer profits by reducing overhead and salaries without sacrificing production capacity. Telework can be seen as merely another form of traditional homework, with the significant difference being that much of it (or at least the parts that are cited as advantageous) is now professional and technical work along with low-status "sweating" work.[25] Numerous companies are finding that the cost of the necessary technology, training and maintenance does not offset the lower salaries and overhead, as expected.[26] Many technologically based home businesses are no longer viable given the shifts in the industry. As Maggie, an IT journalist who has been covering the IT field since the 1980s, tells me, "I think there are some huge changes going on in the industry right now. I also think there are fewer people working at home alone now. A lot of the early webgrrls, and this is just the industry in general, worked in basement shops as Web masters when the Web first was becoming popular. Lots of small businesses would pay a flat fee or some miserable fee to do a Web site. There were a lot more one- and two-person shops. There still are. But I think there are fewer because it's a very tough way to survive." Particular occupations have become less attractive as choices because of their economic obstacles. The output of time looking for new work as a result of the competitive virtual telework industry is significant. Advantages gleaned from telework have been offset by greater demands on time, as well as economic uncertainty.

Rather than bringing radical change to the workplace, new technologies often allow the same old worker relations to be packaged in new metal boxes, and merely facilitates traditional functions of workplaces without benefit to workers.[27] It should go without saying at this point in the book that structures of social relations are fundamental to organizing work practices and thus to the development and implementation of technological objects and processes. As we have seen, rather than taking a technological determinist view that holds that technology itself defines workplace relations, it is more accurate to examine how material practices both within workplaces and in society as a whole engage with technology to produce a variety of effects. Rather than technology inherently determining its own use in isolation from context, decisions about the implementation of specific technologies are influenced or even determined by social institutions, strategic factors and relations of power and privilege. There are many other potential extra-technological reasons for implementation of particular objects and practices in certain ways such as augmented control over information, since in using technology to facilitate work arrangements, employers have traditionally built on existing office organization.[28] In other words, technology tends to be used to reproduce the way business gets done, not challenge it. Administrative reorganizations and restructurings have more to do with workplace relations and politics than the technology itself.

However, the technology itself has indeed facilitated particular kinds of changes. With improved communications technologies, workers can now be made available twenty-four hours a day, by e-mail, pager, cell phone or fax. There is no space that is truly private. The electronic invasion is near-complete. My phone provides me with call display so that when I work at home, I can ignore unwanted callers and continue to work. However, this same call display screen is used by the phone company to send me advertisements. As I type this, the display screen reads, "Do you want to order a pizza or call for a taxi? Press More!" In many cases the technology expands the employer's ability to make demands on work time, translated into "increasing productivity." If the network goes down in the middle of the night, the worker is likely to be woken up by the beeping of the pager demanding immediate attention. Workspace boundaries become fluid; worktime expands. Research demonstrates that for teleworkers, increased productivity actually means, in practice, more working hours.[29] As we saw in chapter 1, the omnipresence of paid and unpaid work demands and expectations, and the lack of boundaries between work and leisure space, can mean that women's labour appears never-ending.

Technologically facilitated communication can also increase expectations of the speed at which work must be done. A downed server must be resuscitated right away; interrupted networks are not permitted to wait until the next morning to resume their flow of data. Homeworkers have always experienced the problem of the blurring of homework boundaries, but telework provides some novel ways in which this can be enacted. A teleworker who uses the home computer for leisure time may feel guilty about doing so when there is work to be done. E-mail may be sent and received at any time. There may be no physical space, as in an office building, which frames the work performed and demarcates the boundaries of work and play. Technologically facilitated homework may simply increase the magnitude of competing demands for women workers' time.

This invasion and level of expectation is often justified on the grounds that the arrangement increases worker productivity. "Productivity" can be a somewhat nebulous concept, referring to work output as well as the amount of time spent by workers. Nevertheless, productivity is assumed by employers and policy-makers to be improved by technologies. This assumption is not without a basis in reality. The implementation of information technologies into work practices can create surplus time for a worker. For example, instead of me physically travelling to the library, climbing the stairs to the book stacks, hunting for and retrieving a journal, then poring through it for the material I need, I might be able to read this journal on-line. I have now saved potentially hours of time. Indeed, it is technology in general that has provided humans with opportunities for leisure and freedom, which a labouring pre-industrial peasantry could have never imagined.

However, in the present context of technological capitalism, this surplus time is not used for leisure activities. Because I've saved this time, and because I can work so much more quickly, the expectation of greater production increases. Work can be done more rapidly, therefore we should do more work. If I can monitor the world's financial markets in real time on my computer, then my employer may expect me to be up at three a.m. to catch the opening bell for the London Stock Exchange. Criteria for what is "productive" consistently intensify in their demands. However, many of these claims of improved productivity have yet to be substantiated, since they are often based on subjective claims of qualities such as "effectiveness."[30] "It is possible," writes Ralph Westfall, "that productivity increases reported ... reflect, at least in part, inflated perceptions or artifacts of the telecommuting situation."[31] In other words, the assumption that IT will necessarily lead to increased productivity is false. There is even a "productivity paradox," which holds that "despite massive accumulated and rising investments in IT, on the whole these have not contributed to significant rises in productivity."[32]

The Gendering of Domestic Time and Space

All shifts in the organization of work are shaped by social structures and relations. As we saw in chapter 1, for example, the arrangement of domestic space along gendered lines means that men and women are likely to experience home-based telework differently. Women's relationship to the home, because of systemic social relations as well as commonplace ideologies that shape the division of labour, is different than men's. Thus any discussion of home-based telework necessitates a "gender lens." Gendered assumptions about women's homework shape women's experiences with this employment practice.

Many forms of homework are grounded in assumptions about women's unpaid labour in the home, which is termed "neo-familialism." Neo-familialism holds that "the family's role in reproduction can be extended into areas previously located in the market sphere."[33] The family in this model is reconceived as a new kind of work entity, in which alternatives to previously publicly available services are provided: care of the infirm and elderly or support of young people who are unable to live independently, for example. While this is often lauded as a "return to self-reliance," in more practical terms it signifies the downloading of unpaid care and service tasks on to the family. This added burden is inevitably borne disproportionately by women. Given this context of neo-familialism, we see that the application of homeworking is intrinsically gendered. Women who perform homework face expectations that they will perform unpaid domestic work since they are at home anyway, the notion that their work at home counts less than work done in the public sphere, and the difficulty of negotiating "workspace" in an arena where they are not expected to have "a space of their own." For women, working at home does not tend to result in more control over their surroundings and activities. In fact, it may even further reduce what little claim they originally had to this autonomy.[34] Thus, the development and experience of telework jobs is fundamentally predicated on who will perform them.

*

Lilith, whom we met earlier, has experienced time invasions throughout her career as a systems engineer and network architect. She recalls her years of working in an "always-available" time frame, both at home and at the office. "When your business phone rings after hours, you're tempted to answer it. Just one more call … and then you're stuck again, working till nine p.m." The central problem, she says, is that customers and employers come to expect that technical workers are on-call twenty-four hours a day, regardless of how (un)important the work-related matter is. "It seems only reasonable," she tells me, "to assume that when you finish or leave work, you can have some time to yourself." I agree that this does seem reasonable. "But if I let my cell phone ring to voicemail after hours and didn't return the call for ten minutes, we would have a big meeting the next day about customer service." Also, she says, in-person meetings or conference calls are another way that workers' time can be unnecessarily consumed. "Many of our meetings went on until two or three a.m. I would try to hurry things along by keeping meetings on topic, but the discussion just sort of meandered from one point to another with no end in sight. When you're sitting on a conference call at ten p.m., after seven hours of a planning meeting, you don't want to chat about somebody's ex–co-worker who did this really interesting thing. You want to go to sleep." I ask her how her employer felt about all of this. Lilith is grim.

"I can't begin to count the number of times I was yelled at — literally yelled at — for not meeting unreasonable expectations." This mentality, she says, was very common in the IT sector in the 1990s.

Lilith tells me another story about missing her birthday party because of work. For her twenty-sixth birthday, her friends threw her a big party. About fifty people showed up to her house to enjoy a barbecue, live music and a bonfire. Half an hour after the first guests arrived, Lilith got an urgent page from the network technicians in Santa Clara, California. The firewall had died. "So," remembers Lilith, "I spent about four hours troubleshooting the problem, on the phone with California and logged into several different boxes. I found out that we had dead hardware that needed to be replaced, and nobody who knew how to replace it was around." The company booked her on a six a.m. flight for the next morning. "While I was working on the problem," recalls Lilith sadly, "my birthday party was going on around me. Then I had to go to bed early so I could get up at three a.m. to get on a six a.m. flight and fly for seven hours to do an hour's worth of work and then fly back that night."

Eventually Lilith quit and moved to another company who believed firmly in working moderate hours. She tells me that at first, she wasn't even sure what to do with what seemed like boundless free time. Leaving work at six p.m. seemed to her like an unimaginable luxury.

*

While IT has provided new types of demands on time and new invasions into workers' space, technologically facilitated changes in the time of work practices vary in their implementation. Rather than the technology intrinsically being responsible for shifts in worktime, work practices and expectations of the employer remain the primary factor in determining how worktime will be experienced. While some women in IT are lucky to work for companies that emphasize shorter work weeks, other women work long hours, either because of the demands of their employer or because they are struggling to combine various sources of income. Among the group of women I interviewed, it was rare for them to work for companies that encouraged shorter work days and much more common for them to work longer hours, including weekends and evenings, particularly if they combined onsite work with freelance or self-employment. "I do about nine hours [a day] on average at work, and then I get home and then I do maybe another four if I'm lucky," said one Web designer. Although Jane tells me that she doesn't believe in spending long hours at the office, she still spends plenty of hours working. "I work something like eight-thirty to five-thirty, and then I take my computer home. I do reading on-line, I go to industry events after work, and then on weekends I get caught up on reading for work." Thinking about it a bit more, she added, "I haven't been on a vacation since January." Our interview is in December.

Mothers of children were also likely to work longer, or more irregular, hours because of the demands of accommodating children's schedules. Mary, who has one child, tells me about her schedule: "I get up at six, work at my stuff till eight-thirty when my son wakes up, get him dressed and fed by nine-thirty, and I drive him over to the babysitter's, then go back to my work. I try to work till five-thirty. I do try to do as much as possible during business hours, and I think it works out to about nine hours a day, though I also work weekends." Laurie, whom we met earlier, is a full-time student and single mom who has gone back to school for an IT degree after several years of working in a technical trade field. She recalls the winter school term ruefully: "I had no child care, so I had drive to my parents' in the east end, then drive an hour and a half to the university, and then back at the end of the day. With all that bad winter weather, I was getting to bed by one a.m. My son is a special needs child so we had to be up early so my son could catch his bus." Neetha, who has two children, is also looking bleary-eyed. "You have these enormous pressures to work long hours. And to be any good at your job, you have to come back and study, and work again. All these six years I've been working full-time, I've been looking after my family, and I've been studying. I come home every evening, and then put another several hours of intense study in. So where does that leave time for family, and kids, and all?"

Small IT companies were especially prone to demanding a substantial time commitment from their employees, and another respondent recalled "staying over at the office, people sleeping under their desks." During her stint at one company, Lilith actually brought a futon to work and stored it in the broom closet, so that she could nap as necessary during gruelling and frequent eighteen- to twenty-hour workdays. Faith recalls that when she worked for a dot-com, six months in advance of having to deliver their Web site, "We were just there from eight a.m. to eleven or twelve p.m., and then on weekends. I finally started bringing my dog to work so that he wouldn't have to be alone so much, and I could take him out." These kinds of work hours and experiences seemed to be a result of increased employer expectations, combined income activities and common practices in the IT industry.

Laurie says that because she is a single mother, she isn't seen as a viable candidate for many IT jobs, despite her qualifications and experience. "There seems to be an awful need for someone who can work twenty-four-seven and fly out of town on a moment's notice." Interestingly, her ability to work virtually is seen as irrelevant. "It doesn't matter that you're hooked up at home, that you have state of the art stuff at home, and can do anything you can do in the office." Employers, she says, despite endorsement of virtual work, tend to prefer the physical bodies of their workers to operate on their chosen time schedule and within their chosen spatial parameters.

Laurie also notes a paradox in the focus on speed combined with endurance: the difficulty and intensity of the most demanding and senior-level jobs preclude the participation of all but the most manic fifty-year-olds. "You're expected as the more senior professional to be putting in sixty, eighty, ninety hours a week, or there's something wrong with you. But when you're fifty, you think, I don't want to do this any more. Yet the work is so complex you need the senior people to do it. It's very detailed. Younger people don't catch the details or have the experience or expertise." I ask her why she thinks companies operate this way when it is clearly detrimental to their workers' well-being. Laurie is characteristically candid, and responds, "We're driven by competing in the global marketplace and corporations' greed for profit."

But not all women in IT are partial to longer hours, and many have decided that they will only work longer hours if forced to. Naomi is skeptical of the value given by employers to long hours. "I'd be suspicious if I had a department tell me they work late hours. Something's out of whack, someone's trying to impress someone, someone's not getting the work done. In most workplaces, unless you're in med school, you have no business being in work that long." Maria, who is in her fifties, has realized from experience that she will burn out if forced to push too hard. "Physically and mentally I can't do more than forty hours and still be healthy," she says. In the past, she worked up to sixty hours a week. She's had to be "ruthless" in sticking to forty-hour weeks, since there is immense pressure to do much more, from the general climate of the IT culture, as well as from other co-workers. She credits her age, experience and a good "reality check" for her determination to balance her life and her work. "You can kid yourself quite a way about being superwoman, and think you can do everything," she admits, "but you have to accept your own limits."

*

Changes in the space in which work is performed are often thought to lead to changes in the structure of work itself. One way in which this was expected to happen was that telework, and by extension "the virtual organization," would lead to "flatter organizations" and a decrease in hierarchies. However, from the 1950s through the 1970s, computerization emerged in tandem with an increase in the number of middle-management positions. Existing power relations were reproduced, not reduced.[35] Thus, while work may be spatially decentralized, it remains conceptually and structurally organized along familiar hierarchies. People may be working in teams, but those teams aren't necessarily equal. Just like when we were kids and picking teams for gym class, teamwork still has a pecking order.

The slightly anarchic, ad hoc method of work organization that was common in the early days of IT work has been appropriated by employers. It is common practice now in the IT field to organize working groups into teams. This is often done for the purpose of "employee empowerment" or "participative management." In practice, teamwork has numerous advantages for workers. People can bring their individual skills, talents and perspectives to the table. Each worker can feel valued as a unique contributor and exercise some degree of autonomy in their labours. Well-rounded projects and people can emerge from this type of synthesis and collaboration. However, teamwork can also be used as a means of social organization, control and surveillance. Workers who express discontent about late hours or the direction of the project may be exhorted to be team players. The employer need not be present at the team table; workers can begin to monitor their own actions and behaviour under the pressure of group conformity. For women, who are not always uncritically accepted as team members, this pressure can be particularly acute. Already potentially excluded because of their gender and consequently their perceived abilities, women in IT often have to work harder to be incorporated into the group as bona fide community participants. This presents them with a troubling conundrum: either they accept the logic of the group, work long hours, espouse shared values and, potentially, become "one of the boys," or they reject the push towards group cohesion and risk rejection. Either way, there is a price to pay. Women who become "one of the boys" often discover that it means erasing any sign of gender difference and playing along in a culture that is often sexist. Women who choose their own path or place limits on their participation may find that their opportunities and advancement are subtly (or not so subtly) blocked.

*

Faith is excited by her work in IT because it allows her to work in a team. She likes the way that putting together a technical project becomes a group exercise, where ideally people have respect for other people's roles. "What keeps me here in this job is smart people," she says, "and getting to work with and learn from them." However, she is less excited by the behaviour of some team members who try to use familiar hierarchies of gender and perceived expertise to their advantage. She finds that many male "techies" can be patronizing and difficult to work with productively. "The designer I'm supposed to be working with is very 'Oh, don't worry your pretty head about that, it's my job.'" She worries that this superior attitude of technical ability conceals incompetence. "He just pretended he was technical. He's been able to cruise within the department to build something and get away with incompetent work, because he's worn this technology hat. Everyone thinks he's technical, but he's not." In fact, she tells me, his success is probably due to the presumed link between gender and technical competence.

While Faith doesn't feel she's faced overt discrimination because of her gender or her visibly East Asian background, and in fact has always felt that she was, more or less, "part of the team," she notes that there are subtle forms of social organization that privilege relationships between males in the workplace. "There are guy things, golf tournaments to go to, that the guys do. I don't play golf so I'm excluded, and it might be mostly social interaction but we know that business gets done that way." Diane, a Web designer, agrees. Her male co-workers play video games together instead of golf, but the same rules apply. "The guys are devoted to this one video game, and they play it like crazy. It's not something that's interesting to me at all, but that was their bonding. Later, you find out about important things they talked about when playing, and decisions that were made." A third woman, Charmaine, who manages Internet initiatives for her company, tells me simply, "I haven't been invited to all things, let's say. And when I've started to talk about technology in front of some male co-workers, I've noticed that they're just not listening to me."

*

Over drinks at a local bar, Cynthia narrates her autobiography to me. She is a thirtysomething woman from Barbados, whose jovial manner belies her on-the-job persona which, she says, is as "the bitch." I raise my eyebrows at this and she explains. "I'm the only female manager. They call me the bitch. I have to be, because I gotta let them know, hey, I'm sorry but y'know, mine is just as big as yours." I ask her how this approach works out. She flashes me a wide grin, showing an impish faceful of teeth. "I tell 'em that every day. And I get the respect that I need, that I should have." I ask Cynthia if it was difficult for her at first. She is fairly young-looking, female, black and speaks with a Bajan accent, all things that might count against her in the typically white North American male culture of IT. She says yes and explains: "With my staff, the guys, I had problems. They felt, 'Oh she's a woman, I don't worry about her.' So when I started giving disciplinary action, and saying, 'Look, I'll fire you if you want me to,' people started going, 'Oh, she's *serious.*'" I like Cynthia, but I get the distinct sense that I should not mess with her.

Back in her home country, Cynthia started out wanting to be an accountant. Driven and ambitious, she was anxious after completing school while she waited for the results of her exams. Eventually, she annoyed her mother so much with her fretting that her mother enrolled her in a computer class to distract her. Shortly after completing the computer class, Cynthia changed her mind about being an accountant. She liked machine language and learned various other skills over the years: DOS, C, Java, HTML, network administration and so on. In her mid-twenties, she decided to start her own consulting business for hardware and software and taught herself the rest of what she needed to know. Two years later, she arrived in Canada.

Her tenacity, skills and ambition earned her a job as a customer support manager in a technical firm. But it wasn't easy — partly, she says, because of employer prejudice and assumptions about her abilities as a black immigrant woman. She has to work harder to earn her authority and employer perceptions of competence, and she feels that employers, even if not overtly racist or sexist, are just happier hiring people who they stereotypically associate with technical aptitude. "If the guy in the interview is looking for a hardware person, he might choose the male candidate, because he thinks the guy is willing to crawl underneath the table, and stuff like that. They think the guy doesn't have to go home and see the kids." Despite her management position and extensive skill set, she does not talk about the fact that she is a lesbian, because she worries about the negative impact that it could have on her already tenuous status. "I'm in a management position, so I have to be very careful. The company is an extremely liberal company, but it's just the nature of the business." I ask about differences between Canada and Barbados in this regard. "I was not out in Barbados. It's illegal in the Caribbean. It's not enforced in Barbados, but it is enforced in Jamaica, and it is enforced in Trinidad. In Jamaica there's a good chance they may kill you." Many of her co-workers have guessed already, but she prefers not to be openly queer at her workplace. That's just how things go, she says.

Diane also finds herself wondering about how much of the exclusion she experiences in the workplace is related to her sexual orientation. At every new job, she has to decide whether or not to come out to her co-workers. "It was frightening at my first job," she says. "I've tried hard not to make it an issue, but at the same time my partner has a female name and that's how it is." She feels that people in IT are more receptive to diversity than in other industries, because many of the workers are young, and "geeks are sort of outcasts" themselves. In another, more traditional field, "like Dofasco," things would not be so easy. In general, she finds her co-workers accept her. She worries, though, that some of the subtler group dynamics in her workplace indicate a discomfort with her orientation. "I'm always asking myself, are they doing this because I'm gay?"

OUTSOURCING: NEO-COLONIALISM AND SWEATING

Outsourcing is the practice of contracting out work in which a company does not specialize (for example, accounting or data manipulation). It allows companies to procure workers for particular tasks while minimizing costs of labour and overhead like office space. It is commonplace for affluent North American countries to outsource IT tasks to geographic locations where labour costs and standards are lower. In Canada, various economically depressed locales, like the Maritimes, have also emerged as centres for this type of work. Outsourced work can be distinguished from contract work in that contract work tends to be done by someone who is assumed to be a skilled specialist and who is retained for specific jobs. Outsourcing tends to be synonymous with subcontracting, where the primary objective above all others is cost saving, and the secondary objective is distancing the company from the day-to-day messiness of governing employees in this area. Outsourcing is a deliberate strategic choice by a company to cut labour costs and responsibility.

Given its global and transnational nature, outsourcing is emerging as a form of technologically facilitated colonialism. It depends on and exploits the wealth inequality between richer and poorer nations and regions, mining disadvantaged areas for their natural resources of low-paid labour. Public relations material for one outsourcing firm states:

> Because of depressed economies, many of these countries [where the company locates itself] have an abundance of highly skilled, under-employed workers who will gladly work for less than a third of the salary for the same job in the United States. Offshore contact centers provide attractive jobs, stimulate the depressed economies, and provide low-cost, high quality services to the U.S.[36]

By this logic, generous First World companies are the saviours of downtrodden Third World economies, dispensing "trickle-down" treats to the needy masses. State sponsorship and subsidization collude with corporate interests to enable this, on the basis that these create jobs and the benefits of capital investment will eventually be distributed downwards into local economies. Underlying this notion is the assumption that "'modern development' [is] the panacea to unproductive backwardness."[37] Local governments fund job training, give income tax concessions and subsidize the already low-paid employment offered in order to reap the benefits of this technologically advanced work. The primacy of capital in setting the agenda for "development" is never questioned, nor is its responsibility towards its "highly skilled, under-employed" workers interrogated, for the relationship is predicated on the assumed legitimacy of technological work to bring economic progress and social enlightenment. Yet there is no evidence that this will necessarily be the case.[38] Indeed, it appears that the more effusive and florid the rhetoric about the ability of technology to benefit marginalized people, the more scanty the evidence that actually supports this assertion.

One way in which the colonization of technical work is enacted is by putting offshore call-centre workers through cultural training, which extends to the creation of a virtual identity for geographically and culturally marginalized workers. Troublesome markers of difference such as accents are eliminated. Sitel's PR material puts it this way:

> Some companies fear that people on the phone in accent laden countries such as India, the Philippines, or Jamaica will either have too heavy an accent to be understood by their American or English customers, or won't be able to relate to them culturally. But with the right training, that can be overcome ... This includes basic lessons on American/UK history, lifestyles, government and language including popular phrases and slang. Once our CSPs [customer service providers] go through the Americanization or UK-centric training, many of them are more easily understood than some regional dialects in the U.S.! We also have American or UK television on in the break room; we play American or UK music and provide American or UK newspapers ... We even give many of our offshore CSPs American or UK nicknames.[39]

Eventually, as they assume an idealized, commercially fabricated identity through instruction and cultural consumption, the virtual workers become more "real" and "authentic" than the actual U.S. or U.K. workers (which presumes, of course, that U.S. and U.K. workers do not have "heavy accents" or use incorrect slang). Colonization entwines with mimesis and fabricates an idealized First World subject from the raw materials of Third World bodies and minds, but the CSP doesn't stop there; she or he becomes a better American than Americans themselves. Fears of the Other are assuaged; the threat of difference is digitally erased. It is assumed that listeners in the U.S. or U.K. will hear an Indian, Filipino or Jamaican voice as incomprehensible, anxiety-provoking, unintelligible, perhaps even menacing, rather than as an echo of home, community or familiarity. The presumed white, "unaccented" middle-class anglophone consumer is differentiated from the actual provider by a distance that is physical, cultural and economic, but the imagined provider is intended to be a comforting replica of a social ideal.

The cover of *Wired* magazine for February 2004 provides another view of the cultural Other that emphasizes Other as different and threatening. The cover shows the face of a South Asian woman as "The New Face of the Silicon Age." Her look is stylized and exoticized. She stares at the viewer with heavily made up Byzantine-icon eyes. A golden ornament hangs in the centre of her forehead, and an ornate gold bracelet encircles the wrist of the hand that covers her mouth. On her palm is a hybrid patois of Java, English, Devanagari script and decorative embellishment, written in henna. Since the hand is placed in front of her mouth, this is a visual metaphor: she speaks in code. Half global citizen, half devouring goddess, she consumes the language of imperial and technological colonization just as she consumes North American jobs. She defies the viewer in her frontal gaze, and the magazine's copy reads, "Tech jobs are fleeing to India faster than ever. You got a problem with that?" And yet, despite her presumed power to gobble up programming jobs, the hand speaks for her. She has no mouth, no voice of her own. She is the imaginary Other.

Donna Haraway noted in her germinal article "A Cyborg Manifesto" that "our best machines are made of sunshine; they are all light and clean because they are nothing but signals, electromagnetic waves, a section of a spectrum ... People are nowhere near so fluid, being both material and opaque."[40] Sitel's PR material echoes this: "We are fully committed to keeping offshore contact completely transparent to the customer." Just as early industrial workers were encouraged to think of themselves as emulating machines in their disciplined regularity, so now are twenty-first-century workers required to mimic the non-space of cyberspace, to be ethereal, incorporeal, merely a Westernized voice smiling down the phone. Visible markers of social, temporal and physical location, of difference and marginality, become invisible. Technology enables a global workforce, but the aim of transnational employment practice is to eradicate diversity, except of course, where it counts: in ensuring low pay and minimal worker protection.[41] When folks said that on the Internet nobody knows you're a dog, is this what they imagined?

Outsourcing looks different in different national contexts. As Carla Freeman, in her study of female teleworkers shows, corporations take advantage of local economies and cultures in the course of implementing their enterprise.[42] In the case of Barbados, companies build on existing ideals of femininity and work practice. In a country where many people work in the fields as a primary source of income, office work carries a relatively higher social currency. "Office girls" are given an elevated social status despite the repetitive dreariness of the work, and they demonstrate this status through their neat, feminine attire.

The transfer of work from high-income areas of the globe to low-income areas is relevant in the context of this book for many reasons. First, it signifies the shifts in IT work that are ongoing. Outsourcing as a practice depends on the global stratification of the labour force, and in many regions, this means that women primarily perform this low-status, low-paid work in the *techiladoras.*[43] It is occupational and industrial segregation on a large scale. The movement of IT work from affluent to less affluent regions also signifies the myth of compensation for skill. The status, value and compensation of work depends not only on who does it, but *where.*

Second, IT work mobility is not limited to call centres. In the 1990s, professional and semi-professional workers in the IT industry in North America could feel that their skills were unique and their labour in short supply. They could feel secure in the knowledge that they were well-paid pioneers in the field. Today, workers in Russia, Argentina and India are demonstrating that the knowledge economy is indeed global. Transnational outsourcing is nibbling around the edges of North American white-collar work. Employers are looking for cheaper pools of IT labour and are discovering that the world is a small place when connected electronically.

In one sense this may render women's labour power more attractive. Women have often been an appealing army of labour when the job calls for lower salaries. Ideas about women's work combine with existing workforce stratification and new fields become feminized. This may mean that women worldwide may have better luck at securing IT jobs, albeit devalued jobs of poorer quality. We are likely to see a renegotiation of work value and the perpetuation of divisions between workers along axes of geography, gender and race/ethnicity. Protectionist currents are emerging as downsized North American workers protest large-scale job loss. For women in North America, this may mean that as they move into better-paying, higher-status jobs in IT, they will see the work accompanied by a drop in its value and status. Interestingly, we may also see a stark division between the U.S. and Canada. Canada is viewed by U.S. companies as an attractive destination for outsourcing. It is clear that labour activism of the twenty-first century will necessitate a global outlook to ensure decent work for all.

*

As I noted in previous chapters, the technoculture of the 1990s was fundamentally a consumerist vision.[44] It was based on the assumed lifestyle of an affluent male IT worker, which included a conspicuous consumption of electronic goods, an abundance of leisure time in which to play with them and a healthy salary with which to purchase them. The technoculture of those who *produced* electronic goods was fundamentally different. Producing the goods that became part of wired fantasies was a much less sexy process. Like the call centres, production tended to involve women working in poorer regions, under difficult conditions, performing repetitive tasks. The current world of consumption *requires* the world of production for its sustenance. The world of IT consumption needs the world of IT production to be invisible, rapid and, above all, cheap. However, the world of consumption has changed. The consumers, now that they have been downsized, scaled back and contracted out, are losing their ability to consume. The formerly libertarian culture of North American IT work is now making very labour-positive noises. As the old saying goes, where you stand on an issue depends on where you sit. Behinds that are no longer in desk chairs are suddenly contemplating walking in picket lines. WashTech, an organization for IT employees on the West Coast, sums up this change in perspective:

> As a high-tech worker, you may not spend a lot of time focusing on workers' rights. You keep your skills up, you learn the latest technologies, and so far you've always been well-compensated and treated fairly ... As high-tech moves from being an infant industry to a mature one, and as the initial dot-com boom becomes a memory, high-tech employers increasingly focus on the bottom line. That translates into pressure on workers — pressure to accept lower wages and longer hours and not to rock the boat ... Maybe you like the idea of a union to help balance the odds.[45]

Alliance@IBM is singing a similar tune:

> Alliance@IBM is made up of career-minded IBM employees who are concerned about our future. We are concerned about recent actions which undermined the retirement security of tens of thousands of IBM's most devoted employees. Over the last several years, we watched as management made reductions in other benefits. It sharply increased our healthcare costs and it excluded many of us from getting paid for overtime. We formed Alliance@IBM to restore policies that value us for our contributions. We are committed to IBM's success, but we are also stakeholders in IBM and we deserve a voice in shaping policies that affect our pensions, healthcare benefits, and livelihood.[46]

The paradox of capitalism is that it contains the seeds of its own demise. Workers cannot produce infinitely. Though employers seek to maximize profits by lowering labour costs, they also need to pay workers enough so that workers can consume the goods that they produce. There is also a struggle in the harnessing of knowledge as capital. Skilled workers are educated workers. Educated workers can read and understand labour codes, hire labour lawyers and agitate for their rights using the very apparatuses of their work. IT labour may be moving all over the world, but wherever employers can reach with their virtual networks so too can union activists. On-line organizing represents an unprecedented opportunity for IT workers. Over the next ten years, there will likely be dramatic shifts in how we understand the global labour process of IT.

Quantum physicists would no doubt agree that time and space are funny, messy, quirky things. Information technologies have had a variety of effects on time and space, and they have *created* new kinds of ways to experience and work within (and apparently outside of) time and space. These experiences are diverse and often contradictory. Yet it is clear that such experiences of the time and space of work do not emerge automatically from IT, but rather from the social relations and processes that are embedded within it.

CHAPTER 5

LOOKING AHEAD

EARLY IN THE PROCESS of doing research for this book, I went to a panel discussion that was sponsored in part by a women's technology organization. The theme was the by-now familiar "women and technology." Several female executives and entrepreneurs in the Toronto IT sector were assembled to speak. The setting was a posh business club in Toronto's tony financial district. The room where we were seated reeked of old money. Self-effacing female attendants spoke in whispers as we checked in. Corporate sponsorship was omnipresent. Sleek male PR droids slithered through the crowd, smoothing the edges of the group, always looking occupied and intent, yet with a veneer of polite formality. The seats were filled with anxious young women, most of them obvious nerdgrrls who would clearly much rather be in a desk chair in front of a glowing screen than sunk into the plush, understatedly tasteful embrace of the fake antique chairs. Everyone drank from the elegantly presented offering of faux crystal and tried desperately not to spill anything on the Persian-print carpet, which, as I noted when I looked down, had its illusion of wealth marred by the obvious seams holding it together. The room was almost full, a turnout of over two hundred people.

The host for the evening, as the event brochure indicated, was a "professional public speaker." He was indeed straight from Motivational Orator Central Casting, with plastered-down dark hair, an understated yet stylish blue suit and well-scrubbed yet faintly plastic skin. He spoke with a measured pace, throwing out tastefully amusing anecdotes casually to the crowd. Our first speaker was a corporate lawyer, who stepped up to the podium to tell us that the provincial government had drafted an intellectual property and privacy act. He was concerned that companies be able to quickly patent and privatize their information, rendering it a tradeable commodity. The personal aspects of privacy, such as worker surveillance in the workplace, surveillance of individuals in the public sphere or the compilation of corporate databases on people's spending habits were not mentioned. This viewpoint, concerned primarily with the acquisition, consumption and protection of corporate information, set the tone for the evening. It also seemed at odds with the spirit of open communication that had first distinguished the Internet and that many of the women I spoke with said had first intrigued them about IT.

The subsequent session on women and technology was primarily a question-and-answer format, with the panel taking questions from the crowd. After the opening remarks from each panel member, audience members submitted questions they would like answered. I settled into my faux velvet chair, pen at the ready, and awaited the pearls of wisdom about women working in IT. The first few questions were banal. How to choose an Internet service provider? How to find out about intellectual property rights? How to make a business plan? The male moderator tried to shift the discussion to women and technology. The crowd quietly rippled with indignation, and the questions kept coming: How to do taxes? How to write-off a small business? Which software was best for intranets? What industry associations would be most profitable? The women on the panel wrinkled their eyebrows in confusion at the change in subject, and the moderator was now visibly discomfited. He tried to rein in the discussion and forcibly shift it back to women and technology. Members of the crowd began to heckle him. Eventually he collapsed, defeated, and allowed the panel to field the influx of mundane business-related questions.

After an hour or so during which the central issue of women and technology was not raised once, I left. As I walked home, I puzzled over this apparent conundrum. Why would a group of over two hundred women, when faced with a discussion that should surely concern them, not wish to talk about their own experiences in technology? Why would they ask small business questions instead of gender equity questions? My puzzlement changed to indignation. What was wrong with them? Why would everyone not want to talk about this? Then

I realized I was placing my own agenda at the centre of my research. I had to take a step back and reframe the question from a strategic standpoint. What choices were being made in that discussion? Why were women finding it more important to talk about tools for small-tech businesses? The answer that dawned on me was that on one level, this *was* about women and technology: women creating a network of information gathering and peer interaction. Women were asking the questions that they needed to ask in order to address their immediate concerns, and they were doing it without fear in a primarily female environment.

If a woman who owned a small business had the opportunity to be in a room with other very successful women, it would be unlikely that she would waste time mulling over the metaphysical questions of gender identity on-line or whether gender shapes technology when she could benefit from the other women's concrete advice. The questions that seemed "unimportant" to me, like how to find a reliable Internet service provider, were significant women-in-technology issues because they *do* affect the quality, success and professionalism of a woman who owns a small business and, by extension, her economic viability. As I realized in the course of the research, women's work in IT has to be situated in a variety of intersecting contexts. The question then became, Why do these particular women make these particular choices in this particular way? I think this question is germane to the issue not only because it allows me to suggest the depth and diversity of people's experiences but also because it positions them as agents who make decisions based on a series of options they feel are available to them. The issue, then, becomes a matter of what options people perceive as being available, and why. Why is a feminist perspective, or even a generally critical one on women's work in technology, an unwelcome one among IT industry rank-and-file? Why do we say that our choices for IT work are so infinite but behave as if these choices are so limited?

Many books and articles about the potential of "cyberfeminism" often focus on the individual pleasures and experiences of cyberspace. There have been some authors, such as Cynthia Cockburn, Heather Menzies and Vandana Shiva, who have emphasized the importance of material relations and women's work in technology and the lifecycle of technological production from its creation by low-paid female factory workers to its disposal in landfills. These have been influential to scholars but rather inconsequential to policy-makers and the IT workers themselves. In a political context that privileges individual achievement to the exclusion of social critique — which regards technological developments as willful yet neutral, as inevitable but not-quite-here-yet, as highly visible and valuable commodities with invisible, insignificant producers and as existing outside of social and economic relations of power and privilege — such a gap is hardly surprising.

Despite my awareness of women's diverse choices, and my generally positive inclination towards IT jobs for women, I think the absence of substantive discussion around women and technology work is telling. I remain troubled by the absence of an IT-industry critique by women's technology activism groups. Crucial day-to-day issues — of access, safety, discrimination, institutionalized inequality and the demands of the IT labour process — are not being fully addressed. The discourse and policy around women's IT work is reduced to two relevant but inadequate areas: skills and opportunity. And yet ample evidence indicates that these two things, while important, are not necessarily the prime determinants of equity for women in the IT field. As we have seen throughout this book, there are numerous structural factors that pre-date IT as a field of employment and that continue to inform existing workplace relations for women workers. The vision of the technological self remains personal and not political; it is a model of individual information consumption and positions the IT worker as a liberated free agent. There is a sense that all of the nagging, tiresome problems that early female pioneers confronted have been solved, or at least they are not worth worrying about. If gender, race and class exist, they are seen as merely individual artifacts and not relevant in the new economy where, despite downturns, there is opportunity for all.

A report called *Women in the New Economy* confirms the pervasiveness of this perspective. It notes that women are apt to experience gender discrimination but interpret it as other forms of discrimination. For example, women may identify it as poor treatment because they are older or single/working mothers. "It seems the gender issues look different in the new economy," suggests the report. "People 'talk the talk'; however, they excuse inappropriate behavior because everyone is under such stress, or people are so young."[1]

This focus on erasure of identities and social power consequently manifests itself in a perception that these markers of social positioning are not relevant in IT. For example, the January 2001 meeting of DigitalEve, one of the women's technology groups in Toronto, featured the topic "Getting Women Online." The literature handout stated, "The Internet is the great equalizer. It knows no barriers. Geographical barriers. Cultural barriers. Social barriers. Age related barriers. Or gender barriers. On the Internet, we women can all be successful, if we believe we can." This is a tempting viewpoint, but one that does not appear to play out in material relations and experiences. Barriers and systemic obstacles *do* exist and to pretend that they do not, or that they can be overcome simply by positive thinking, does a disservice to women who are then disenchanted with the IT workplace and the climate they encounter. It means that women do not have the language nor the tools to talk through and confront gender-based problems they may find there.

Indeed, women are often hampered by the rhetoric of individual opportunity. Since women who enjoy IT are often viewed as "unusual" or "non-traditional" in the first place, it is quite easy for them to feel isolated and alone when they experience difficulty in an individualistic climate. In a culture where success is seen as a direct result of human capital investment, failure can be viewed as a matter of individual fault. Women begin to feel that their difficulties are simply a result of not believing in themselves. Perhaps they think that if they tried harder or were more of a team player, they could get ahead. Rather than assuming everything is okay, we need to have an open and frank discussion about the pervasive stratification of the IT workforce along gendered lines.

We also need to critically examine the role that IT corporations and sympathetic state bodies play in defining the criteria for what issues are important. When doing the research for this book over the course of several years, I heard a lot from institutional IT sources, such as corporations, the government and IT women's groups, about the need for "lifelong learning," "entrepreneurship," "flexible work arrangements" and "skills upgrading." I did not hear much about how the IT industries and corporations themselves have a hand in creating and perpetuating many of the problems that women in IT face. Women told me that they felt ashamed of their griping, of what they perceived as their technical ignorance and for complaining about poor working conditions in an economic climate where they felt they should be grateful for having a job. The flip side of the individualist ideal of self-empowerment is self-blame. And yet job instability and poor working conditions in the IT industry have little to do with individual workers. Rather, it is often the desperate drive for short-term profit and the greed of governing executives that drive the actions of IT corporations such as Nortel. Nortel benefited from the telecommunications boom in the 1990s, particularly in the United States, where numerous telecom and Internet startup companies proliferated. Nortel did a brisk business selling equipment to these small startups. Yet also as part of the boom, startups borrowed heavily on credit and invested in hype. Sadly, hype doesn't pay the rent as well as cash, and scores of startups folded. As a result, the cheques to Nortel were not in the mail.

As of the writing of this book, 60,000 employees had lost their jobs at Nortel. The company's accounting records as far back as 2001 are under review for falsely documenting the amount and the date of both profits and losses. The U.S. Securities and Exchange Commission issued a formal order of investigation. The Ontario Securities Commission and the RCMP are making "informal inquiries," and the U.S. Attorney's Office in Texas has begun a criminal investigation. Numerous class-action lawsuits by shareholders are pending. Various euphemisms, such as "accounting irregularities," "accounting snag" and "erroneous reporting of quarterly results" have been used to describe the situation that Nortel is presently in. *The Financial Post* is not so generous, terming Nortel's multimillion-dollar profit overstatement "a sham, the product of loose accounting and insufficient oversight ... the firm's finance staff broke accounting rules in an all-out effort to hit profitability ... and to create the appearance of a miraculous turnaround."[2] In 2003 and 2004, at the same time that the "accounting irregularities" were emerging, Nortel executives such as Frank Dunn enjoyed bonus payments of up to $2.15 million. As part of the bonus scheme developed in early 2003, executives would receive these payouts even if "the company reported huge losses as measured by generally accepted accounting principles."[3] In other words, not only did executives get paid a salary to do the job they were hired to do, they felt that they deserved special treatment even if they bungled it. Now *that's* self-empowerment. The story of Nortel is by no means unique in the IT field. Other companies, such as WorldCom and Lucent, likewise used the hype and poor judgement of the IT boom to reap fabricated profits at the expense of shareholders and employees.

The IT industry has not been immune to the increasing structural instability that characterizes work in other fields. Work is increasingly casualized and unstable for growing numbers of workers. However, what is unique about the IT field is the notion, which persists despite a great deal of evidence to the contrary, that *this form of work is different and better. This is progress.* Indeed, with the development of IT and the evolution of the IT field beyond the nuts-and-bolts world of technical work, many new positions and occupations have been produced. Many of these new kinds of jobs are especially attractive to women, as they are built on existing fields in which women are likely to be found. Jobs such as information architect, Web designer or content manager were nearly unheard of a decade ago. IT can be, and is, a challenging field for women where they can succeed. Women have shown themselves to be extremely competent in handling all elements of the IT field, from creation to implementation to use of the technologies. They often enjoy the work and feel comfortable in the IT environment. I am inclined to look for the positive developments that IT has brought to women's employment possibilities. I find many. This is progress, in some ways.

Unfortunately, in all kinds of technologically based work, women continue to confront structural social relations that organize their participation and experiences in the workforce. They continue to experience industrial and occupational segregation as well as wage disparities, based on gender as well as other intersecting social locations. They continue to struggle to articulate the contradictions between expectations and responsibilities of domestic labour, as well as conflicting messages about their intended role in the paid labour market. They continue to be under-represented in positions of power, both in terms of workplace administration and organized labour. Technology itself, despite how wonderful, interesting and useful it may be, does not change workplace or social relations. Work continues to be informed by structures of power and privilege, by access to resources based on systemic relations and by historic traditions of practice and organization. The fact that IT *could be* and is such wonderful work for many of the women I spoke with is what makes addressing these issues particularly crucial. Nobody wants to argue in favour of women taking on jobs that are soul-crushing, mindless or destructive. It is the incredible potential of IT as a set of professions for women that compels me to critique the existing conditions that impede women's full incorporation into the field.

In 2004 the need for feminist voices in the IT industry is critical. Global outsourcing, work uncertainty and erratic employment arrangements are crawling up the job ladder. Pink-collar IT job ghettos are becoming entrenched worldwide. Women's over-representation in the lower job strata is worryingly consistent despite some major gains and the emergence of important female figureheads. Many groups of women are still struggling for access and substantive equity. And yet, among many advocacy groups for women working in IT, the silence on this front is deafening. I do not mean to dismiss the good intentions and hard efforts of those seeking to improve women's work conditions in IT. Many thousands of low-paid or unpaid volunteer hours have been expended by dedicated women who have a genuine interest in making the IT field a better place for women. However, I am concerned that there is no substantive critique of the industry nor of the IT workplace itself. This is where much of the responsibility lies for creating a good quality working life. And this is where many of the structural inequalities are reproduced.

I do not believe that IT itself creates certain kinds of working conditions. People with particular interests create working conditions. IT is simply a medium or process through which work is experienced. It suggests possibilities; it does not determine practice. We need to look beyond the technology as a set of objects, to the people who create and use it, and ask what we can do to make work better and more fulfilling for all workers.

Women's Views on IT Activism

Neetha and I meet in a Tim Hortons in the north end of the city, in a neighbourhood characterized by low-rent concrete monoliths and strip malls. It is a cold winter night, and tiny needles of ice pelt the windows that are glazed inside with a warm mist from coffee vapour and the aroma of donuts. We snuggle into our winter coats and nurse our steaming mugs of Tim's special blend. It's a quintessentially Canadian evening and a stark contrast to India, Neetha's country of birth.

After completing a bachelor's degree in mathematics and a master's degree in economics, Neetha worked at a bank in India for fifteen years. As a bank officer she became interested in the social aspect of her job, particularly initiatives such as small business loans to women and funding social work. At that time, she says, computers were just being introduced to the financial sector. Neetha saw the potential of these new machines and quickly computerized the branch. She is nostalgic as she recalls her first encounter with computers: "I had a computer when it was very new at that time, and then I learned, what was it, Wordstar we used to call it, I learned Wordstar and Lotus on that. It was such a big thrilling thing, you know. Computers are such fascinating things for anybody … it's just the ease with which you can do work. And the ease with which you can write. It's just amazing." She immediately sensed the potential for making work easier. "In the bank I would see the tremendous difference it made, because we used to do everything manually, all the books, and all the balancing, and all the filing, and it would take a lot of time …You have the computer and all the work gets done. It has this tremendous benefit of just wiping away all kinds of problems." She also discovered benefits for her own work style. "I was always too lazy to write. I stopped writing with pen and paper, because I never finished anything. With a computer at home I could do it, especially when I could go back and edit. And then the most recent thing which I've really found is the communication, like the e-mail thing … it helps to create a group and bonds people together and you get a lot done." She is passionate in her declaration: "I'm in love with computers!"

Where she works now in Toronto is a far cry from the hierarchical, structured workplace of an Indian bank. Building on her interest in women's issues and her desire to work in the social service field, she works at a shelter for abused women, which is run by a feminist collective. At first glance this might not seem like a natural habitat for a computer-loving number-cruncher, but she is excited by the possibilities that computer technology offers to the shelter. Just as in the bank, shelter workers are finding that computers can make their work easier and can help them to expand their services. For example, says Neetha, she is now able to create and keep a database of all the women who have used the shelter's services, in order to assess the effectiveness of their programs. "What good are we doing in the shelter? If you can prove over a long period of time, like ten or fifteen years, that the families that come out of the shelter end up being stable people that don't turn into criminals and don't turn into a waste for a society, then we've done our work, we've brought some value to society. That kind of tracking we haven't been able to do because it's just so hard. Women leave, it's hard to keep in touch with them, you don't have a database and you don't follow it up, so we're hoping the computer will make all these tasks easier for us."

Another initiative that Neetha is spearheading is the shelter's Web site, which will provide resources and information not just for local women but for women worldwide. She envisions an on-line shelter and women's services network that would be able to help women and other service providers locate the resources and assistance they need. Social and geographic barriers can be mediated, and more women in need can be assisted, with the careful implementation of the appropriate technologies. However, technologies are only as good as users' access to them. Neetha is worried about this. The women she sees on a daily basis may have left their homes in a hurry, with only the clothes on their backs. They may be from marginalized social groups who do not have many resources on which to draw. They may be new to Canada and their English skills rudimentary. In any case, it is very likely that a high-speed Internet connection is one of their less immediate concerns, and an act as simple as checking a Web site might be out of reach. Neetha frets that this division, between people who have minimal or no access to IT and people whose computers are as close and familiar as their refrigerators, will further exacerbate social and economic inequality. Women who are unable to develop these skills may be left behind in a job market that privileges technological aptitude. Access is only the first step on the road to mastery of technology, but it is a critical one. People cannot become controllers or creators if they must struggle even to become users. As Ursula Franklin points out, admittance to a community of knowers is of vital importance to gaining power in technological fields.

I ask Neetha how she positions herself relative to feminism. Her answer is unambiguous. "I'm a feminist, absolutely." The inequalities that Neetha sees on a daily basis are entwined with issues of power and control, largely in terms of gender but also in terms of other social locations. Because of her work with marginalized women, she is critical of women's technology organizations that do not have a feminist dimension to their work. But also because she experienced technology as an empowering, liberating force, she thinks that access to technological tools and knowledge is critical for all women. She worries that more affluent women, who enjoy easier access to technology and to the resources that enable them to control their IT work, forget that there are still crucial disparities in women's IT experiences, and that it is these privileged women who determine the agenda of activism around technology. Neetha would like to see the debate around women and technology framed more broadly, beyond limited issues of career advancement. "I know there are a lot of women who are trying to make a career in technology, and they come to women's IT groups and they want to advance their careers, and whatever, and I think that's fine, but it's still such a privileged group of women." Iconic women who make it to senior positions in IT corporations, such as Carly Fiorina, the CEO of Hewlett Packard, are certainly role models, but not representative of the experience of low-income or marginalized women or even middle-management women in the IT field. Neetha feels that too many resources are expended on helping the "haves" in IT "have more." She is cynical about the constant focus on women's professional careers, particularly in IT business. She argues that most women's technology advocacy groups such as DigitalEve, WebGrrls, WorldWIT (Women In Technology) and Wired Women "came more from a need for advancing their careers ... rather than from their position as women. It came from a different place ... a lot of the women who are in these groups are really ambitious women whose sense of self is derived from their work." A narrow focus on business networking, she feels, doesn't solve the fundamental problem of technological inequality, nor the challenges and discrimination that women face institutionally.

As I write this, two e-mails from different women's technology groups arrive in my inbox. The first, from Wired Women, advertises a seminar called "Burnout or Balance: The Working Woman's Guide to Easing the Career/Life Crunch." The seminar promises to help me plan proactively for "meltdown prevention" and confront my "personal challenges to creating a more vibrant, balanced life." The second, from WorldWIT, advertises a camp that promises to help women attain their career goals. Sessions include networking, self-branding, self-promotion and "helping us identify the ways in which we drag our hind legs and sabotage the passion held hostage by the way we spend our time." As usual, the message is that it's up to individual workers to solve the problems created by systemic workplace and household demands through their individual choices. In these e-mails, as Neetha also notes in her comments about women's technology groups, there are significant absences in the content of the debates around women's participation in IT. Discussions about women's and feminist issues, such as child care and the demands of domestic work, quality of working life, the wage gap, systemic discrimination in the workplace, government initiatives around pay equity and family policy and the global relations of gendered IT work are either absent, truncated or framed in individualistic terms of self-empowerment and life choices. And yet, as we have seen throughout this book, these structural factors — of policy, workplace culture and gendered norms of paid and unpaid work — play a major role in determining women's decisions and work experiences. But perhaps the most glaring omission of all in these types of material is the subject of the technology itself. No mention at all is made of actually using, accessing, creating or controlling the technology. Virtual organization can be a powerful tool for women, but if women can't access the relatively expensive objects nor the knowledge to use them, then they are prevented from adding their voices to the conversation. Ultimately, they are also excluded from technological autonomy.

*

In contrast to my chat with Neetha in a donut store on a January night, my interview with Gillian is conducted while immersed in a warm sunbeam, as the light streams through the large warehouse windows of a trendy downtown café. We perch gingerly on minimalist chairs and converse over fusion cuisine. Gillian is in her early thirties and works as a project manager for an IT company. She began her career as a receptionist with a lot of spare time to surf the Web, increasing her confidence and Web know-how. Eventually she created the first Web site for her company. She belongs to a women's technology group. Unlike Neetha, she is in fact looking for cheap skills training and business networking. "I think the group does a really good job in helping us network with each other. Educating and having the resources to educate, so the training seminars and stuff like that are very valuable. I often wish there was more I could leverage off that ... I think that those are the two main values, networking and leveraging the expertise off one from the other." It is often hard for me to follow what Gillian is saying, as I am unfamiliar with the jargon of the business world. I am left wondering what concrete advantages "leveraging expertise" actually have for women working in IT.

Initially, she says, she was hesitant to even join a women's IT group. "I was a little intimidated by this all-woman type of organization. I was afraid that it was going to be like a very, um, I guess I figured it was going to be very *left,* extreme left-wing feminist kind of program where I just wouldn't fit in, a very militant type of organization." I am curious about what this means, so I ask her to expand. She seems to have difficulty generating anything specific. "I guess when you see people who are organized, people who develop an organization if there's a certain niche, you expect them to be at the extreme end of that particular niche, whatever that is. I didn't realize that it was all about women communicating with each other. I thought it was going to be like, you know, and this doesn't sound politically correct when I say this, but I thought it was going to be one of those men-bashing kind of left-wing groups that was all about feminism forever." So what brought her to this group in the end? Gillian points to the safer space that an all-female environment can create when asking "stupid" questions about technology. "I think what it was, was that it was an open forum for me to be able to post my questions and discuss my thoughts with a group of people that I knew were not going to condemn me for being a newbie. I could pose the questions and not sound like an idiot."

In speaking to many of the women, I note this interesting inconsistency: they are fearful of asking questions about IT in a mixed-gender space, but yet they are reluctant to identify a need for women's activism. They are able to tell stories about workplace indignities, such as a software saleswoman whose boss habitually grabbed her breast during meetings, but they are unwilling or unable to articulate a sustained critique of systemic sexism. Part of this seems to be derived from the individualist culture of the IT field: it's every (wo)man for herself/ himself. However, after our interviews, during which I tried to be as non-committal as possible about my own perspective on feminism and women's activism, many women told me that nobody had ever spoken to them about this subject before. Simply through telling their stories to a stranger, they began to understand their experiences as women in IT in a different way. This is particularly noteworthy because these women came from women's technology advocacy groups. It seemed from our discussions that despite the apparent attention to "women in IT" as a policy issue, there are few opportunities or spaces that actually allow for a pointed critique of the industry.

Farideh, who spent several years working in the financial sector of the IT industry before she turned to creating multimedia CD-ROMs, expresses this concern bluntly. In the beginning, when women's technology organizations and on-line feminist activism emerged, she says, it was originally an attempt to create and nurture a supportive community. E-mail, bulletin boards and virtual communication in general seemed like a way to generate and spur women's activism. But Farideh reports that she is more cynical now. She noticed that a feminist spirit did not proliferate in these communities simply because women were present. "I think over the years we've learned that the medium cannot be the message. Beyond just being virtual communities they have to have clear mandates and clear directions. There are directions, but in ideological terms I'm opposed to those directions … There is a kind of a glorification of the industry and we are great, we are working here, we are working in high tech, isn't this great, here are some companies, let's get to know them … There is no real critique being articulated by the organizations. It doesn't come across in any of the publications that they put out, it's not a subject of discussion at any of the events, and so, it is very middle class, or aspiring middle class. And

I think that's probably what a lot of people use it for. A lot of people use it as a stepping stone, y'know, a way of establishing connections, precisely because it allows them to network and possibly find jobs that they want to find." Farideh joined the women's technology group hoping that it would serve as a political force for substantive change in the IT industry. She was eager to talk about systemic racism and sexism, about poor working conditions enabled by masculinist corporate IT culture. She now feels very disillusioned by women's IT organizations and no longer participates, preferring to direct her energy and activism into her multimedia projects.

Donna, a young, ambitious woman in IT marketing, is very interested in the business model for IT women's organizations. She has volunteered her time to position her organization of choice as an attractive option for IT corporations. Her vision is of an organization that provides female IT workers with networking opportunities, and companies with a captive, relatively demographically homogeneous audience of female IT professionals. Her plans are grandiose and her tenacity and energy for this project evident. Between sips of venti latté at an upscale coffee shop, she explains her blueprint to me. "We can use all our contacts to go for the corporations that have the biggest budgets, that are willing to sponsor organizations or events that are going to help them with their technology, and improving their position in the marketplace: the Microsofts, the IBMs, the big players." She sees women's IT organizations as providing a pool of subjects that corporations can target. Donna suggests that women would benefit as workers, but it seems apparent to me that companies would benefit much more by both creating "disciplined subjects," who learned to speak the language and operate within the mindset of the industry, as well as consumers who would then purchase the products and services that the company offered.

In addition, corporations could teach women how to reproduce these relations. For example, Donna brought in guest speakers to teach the women how to market products to other women on-line. "We developed a full marketing plan, about here's what we want to achieve, certain objectives, we talked about strategic alliances with other groups, we got into various tactics, we set up a newsletter, we set up our own e-mail list. We just came up with tons and tons of ideas, we talked about keeping profiles on our customers so we can do database marketing on-line, and also it was going to help us with getting sponsors ... And these organizations, I think, really need to start working more like that type of a business rather than like a charity type." The role of corporations in this case is envisioned as both descriptive and prescriptive. Because, Donna feels, women are already working in the IT industry, they will be receptive to companies being involved in activist groups. However, the role of these companies in setting the agenda for the group, and defining the terms of engagement with IT industries, is somewhat more disconcerting. This approach focuses heavily on teaching women to be consumers of IT gadgets and goods, and to accept the often problematic mindset of the industry. It is less an introduction to IT than an indoctrination.

One recurring motif I noticed in my conversations with these women was the struggle over the use of particular concepts, such as "empowerment" and "networking." For Neetha, empowerment meant that women had more choices for literal survival and a greater ability to exercise those choices. Access to technology was part of this, but Neetha felt that this access had to be situated in the larger context of women's political and economic concerns. But for other women, particularly women who worked in the business sector of IT, empowerment could mean anything from the ability to consume products to the ability to access one's e-mail more easily. Another contested idea was the role of networking. In Neetha's world, networking could mean technologically facilitated access to a group of people and resources, such as a Web database of women's shelters, that could make the difference between life and death for women in jeopardy. In many progressive IT development programs worldwide, IT is seen as an important tool that can enable women to enhance their economic and social autonomy through building networks of information and community.

Women in Global Science and Technology (WIGSAT) stands out as one group that has produced a rich, global feminist plan of action and sustainable development using IT as a tool and resource. Their concept of networking includes working with NGOs (non-governmental organizations) and multilateral coalitions of women's groups in order to guide policies and projects aimed at concretely assisting women worldwide, such as the "ICT for Rural Women" initiative that provides a network of resources, events and organizations focused on how women can use IT to support their grassroots productive enterprises like subsistence farming. Yet in the discourse of many women's IT groups, "networking" was often simply an exchange of business cards. In the "information revolution," the language of rebellion and social uprising is often used to convey concepts and meanings that are, in reality, anything but. Though the tools and practices of IT work do indeed contain the seeds of empowerment, as demonstrated by WIGSAT, such terms are often reduced by the IT industry to signifying one's choice of gadget consumption.

*

It has been over 150 years since Ada Lovelace developed the conceptual framework for a programming structure. It has been six decades since six women programmed the first electronic computing machine, ENIAC, and half a century since Grace Hopper's development of the first compiler. Though it demanded a great deal of mathematical and conceptual skill, in its early days programming was seen as menial, tedious work compared with the more valuable task of actually constructing computing machines. As such, programming was considered appropriate work for women. Now, only about 23 percent of programmers are female, and on average they make $5,000 a year less than their male counterparts. (See Appendix A for more detailed data on IT industries and occupations.) In the spring 2004, York University hosted a programming competition for high-school programmers. Several photographs of the event were posted on York's Web site. I squinted at the photos for several minutes, scanning the teams of programmers, to see if I could locate any girls. The best I could do was locate the back of a single pink sweatshirt in one image. It seems that these days, women are more likely to be found as "virtual assistants" than programmers. One of the most common IT-related occupations for women is data entry. CodeBaby Corp., based in Edmonton, is working on "every manager's dream worker: a virtual assistant that works 24 hours a day, seven days a week, doesn't ask for vacation, never gets sick, is always pleasant, and looks sharp."[4] According to CodeBaby's CEO, the "friendly human face" of the virtual assistant is "generally female."

Looking Ahead

Women have been involved with technology throughout its history, though they have not always been recognized or admitted to the community of "knowers." Women enjoy working with technology and many pioneers in the field have been and are female. In jobs where IT use is high, an equal majority of men and women (64 percent) say that IT has made their jobs more interesting, although more men than women are likely to say that IT has improved their job security.[5] But as we have seen throughout this book, women workers in IT continue to experience job segregation combined with devaluation of their labour, persistent wage disparities and workplace practices and cultures that inhibit or limit their full participation. As I have emphasized, these familiar trends are particularly interesting given the promise of IT work. Though the development and implementation of IT has altered the content as well as the location of women's work in some ways, in other ways this work has remained stubbornly familiar. IT has facilitated major shifts in the process of work, but as part of this, it has also reproduced patterns of inequality and marginalization. Canadian women's employment in IT work thus represents a tension between new technologies and old ways of doing business. Because of this, the "add women and stir" approach to getting women into IT does not work. Access alone, while a critical first step, is not enough. Without attention to the fundamental problems of working conditions and work quality, women will continue to be marginalized in IT professions.

In terms of women's future IT participation, there are a few promising signs. One key indicator of women's status in IT is their participation in education and training programs. Women's visibility in university applied sciences and engineering fields has increased. In 2000, Canadian women made up 24 percent of university graduates in this field, up from approximately 20 percent in 1996 (in comparison, women make up about 59 percent of all university graduates).[6] Many IT and computer science departments have taken seriously the problem of a culture that does not welcome women's participation. For example, as documented in *Unlocking the Clubhouse: Women in Computing*, as a result of various concrete initiatives aimed at increasing women's role in the field, the percentage of women enrolled in Carnegie Mellon's prestigious computer science program was 42 percent in 2000, up from 8 percent in 1995.[7] Yet there is a so-called leaky pipeline phenomenon, which means that despite women's early interest in technology-related fields, their attrition rate is high.

Another indicator of progress is wages, and the data here are equivocal. Overall, women's wages in most IT-related jobs have shown an upward trend since 1997, although wages for computer and electronics engineers have dropped in the last two years (see chart 5.1 for data on selected IT occupations; also see Appendix A for more detailed data on IT work). The wage gap in some occupations has decreased, and in some fields, such as computer programming, male and female wages are nearly equal. In other IT-related occupations, the wage gap is increasing. Female computer engineers, whose wages once climbed briskly to the same rate as their male colleagues in the late 1990s, saw the wage gap yawn open again in the last few years. In 2003, female computer engineers made just under 83 percent of what men in their jobs did, the same gap as in 1997.

Chart 5.1: **Average hourly wages for women workers in selected IT occupations, 1997–2003** [8]

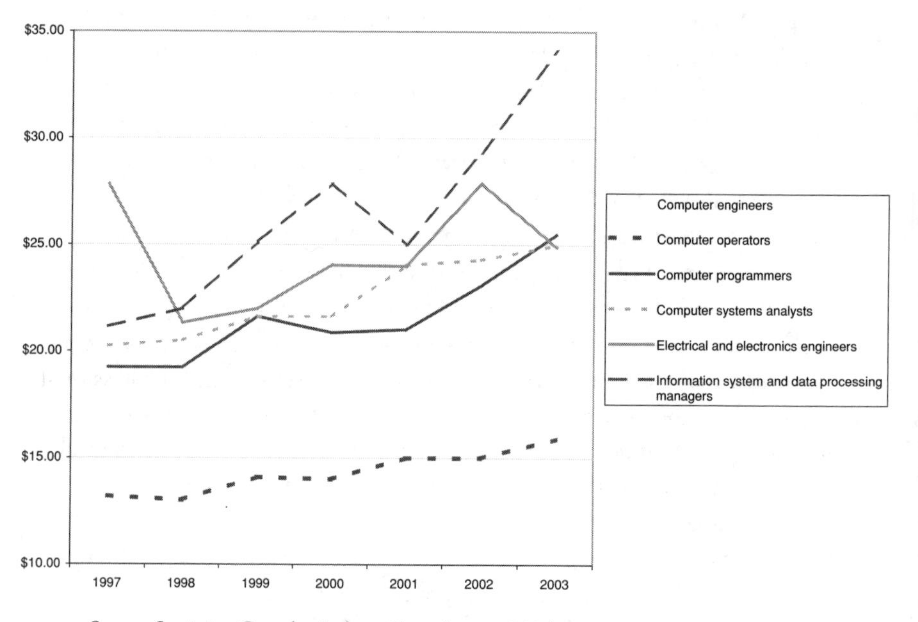

Source: Statistics Canada, *Labour Force Survey, 2004* (custom tabulation).

How can we keep women working productively in IT fields? What has to happen to ensure that women become full partners in the creation, application and control of IT? First, while ensuring women's equitable access to IT is important, access alone is not sufficient. Providing opportunities for women in IT that don't address the substantive barriers to their involvement and career development is like providing access to a person in a wheelchair by installing a staircase. A few tenacious souls will figure out how to make it work, but most will find somewhere else to go. Melanie says that while she makes a point of sharing her interest in technology with elementary and high-school girls, the technology itself does not seem to pique their interest. While they use the technology on a daily basis, they tell her that they are uninterested in making a career of it. "It just doesn't turn them on at all, even though they use it every day. I went and did a presentation a while ago for high-school girls, and they were like, 'We love this site and this site and this site,' and I was like, 'That's so great, okay so how many of you are interested in doing this as a career?' Nobody. And they were like, 'We go to a technical school, and we study C++ and we hate it, it's boring, we never want to see another computer.' So I think it's really more about making the opportunities and the reality of those careers more appealing, rather than making it something like, we have to do this because we're not there."

Second, gender and social location has to be understood as more than a set of individual attributes; it's a major factor in shaping workers' needs, choices and experiences. There are clear differences among groups of workers. For example, immigrant women of colour are disproportionately represented in low-end IT work such as data entry and IT equipment assembly, while women in general are almost entirely absent from the ranks of IT manufacturing and engineering management. Women that do make it to senior and professional positions are consistently underpaid relative to male peers. In terms of employment earnings in IT professions, Canadian-born women of colour fare worst of all (see chart 5.2).

Chart 5.2: **Average annual earnings of computer and information systems professionals by occupation, visible minority status, immigrant status and gender, 2001**[9]

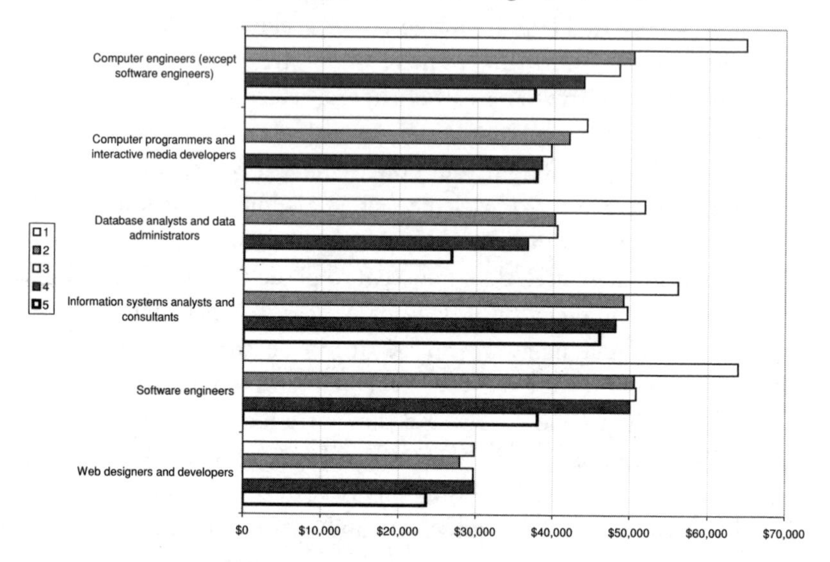

Source: Statistics Canada, *Census of Population 2001*
Legend: 1. Non-visible minority, Canadian-born men; 2. Non-visible minority, immigrant women; 3. Non-visible minority, Canadian-born women; 4. Visible minority, immigrant women; 5. Visible minority, Canadian-born women

---

Diversity in the workplace is about more than celebrating a few extra holidays or adding a few new faces to the staff. It's about putting structures and systems in place that support workers, provide real opportunities for advancement and growth and enhance the quality of life for everyone, not just the women who become "one of the boys" or "play by the rules" of the approved culture. Workplace diversity and equality flourishes in a culture of mutual support and respect. In fact, a study commissioned by the Canadian Policy Research Network found that being treated with respect was the number one item that employees considered "very important" in a job.[10] There are also gender differences in what employees look for in jobs. While both genders listed being treated with respect as their top priority, 81 percent of women felt it was very important, compared with 67 percent of men.

IT should be used to improve, not detract from, worker autonomy, creativity and communication. For instance, IT has increased the potential for flexible work patterns that would enable workers to manage paid and unpaid responsibilities. However, in practice this has tended only to improve the conditions of male professional workers. Thirty-eight percent of women feel that their job has a serious deficit in terms of providing flexibility to manage family responsibilities and, interestingly, this perception grows more acute with the increased professionalism, education levels and seniority levels of the job.[11] A focus on the gendered nature of paid and unpaid work patterns is needed to ensure that both men and women can benefit equally from some of the unique features of IT work.

An example of this principle in practice is Lilith's company. Lilith remembers working years of hundred-hour workweeks for major southern California IT companies in the 1990s. She describes her current position with her current company as the best job she's ever had. Part of what makes this job so great, she says, is that the company culture militates against long hours, and this practice is supported by the CEO all the way down through the management structure. There is almost no overtime. Careful attention to the organization of the workplace is evident. In addition to a supportive management, Lilith points out, to avoid overtime a company needs to have enough employees to do the work within working hours, and thus they need enough money and resources to pay them. While this attention to worker well-being may seem like touchy-feely techno-hippyism, at Lilith's company, it emerges from pragmatic business considerations. "Basically," says Lilith, "it boils down to this: get good people and treat them like human beings." Workers are less stressed and take fewer sick days. Workers with children or other responsibilities can more easily and effectively manage them. Workplace diversity is also about an institutional responsibility to confront discrimination openly, rather than pretending that technology has made social inequality passé. Lilith compares her current company, where a sexual harassment incident was swiftly and considerately dealt with and the perpetrator disciplined, to a previous company where sexual harassment was openly indulged in by management and where data personnel insisted that their computer lab was a "boys club."

Erin, the intranet specialist, shares Lilith's good fortune in working for a progressive organization. She loves her workplace because it has put measures in place to ensure that workers can control their work arrangements and manage their other responsibilities with a degree of flexibility. Her employer allows workers to set their own hours — Erin, for example, prefers to work from ten to six, while an early bird co-worker likes to work from seven to three — and to work from home. Erin thinks that the technology should be used to create more flexible, worker-friendly arrangements. "Some days I get up and I have to have loud music and I just don't feel like coming in, so I work from home. I just e-mail people and let them know where I am, and what I'm doing, and when I'm available ... a lot of these things can be technologically facilitated." Employees manage their child-care demands with various types of solutions, including occasionally bringing their children to work. "People bring their babies in, and that's no big deal," says Erin. "You go into the boardroom and there's some kid on the floor, watching a video or something. It doesn't matter, it's great. This is a very supportive environment for that sort of thing. Which is one of the reasons I'm here." Erin, who doesn't have children herself, still supports a parent-friendly workplace policy as something that takes pressure off women. She feels that a workplace culture that supports employees' other needs benefits everyone.

While Erin and Lilith focus on the big picture of how workplace improvements benefit everyone, it is sadly common for women who have "made it" in male-dominated fields to feel that they have no obligation to help other women. As one veteran of the IT field told me, "I've made it to the top without any help. Why should I help anyone else?" And yet, often the most transgressive and productive thing that women (especially the few women in positions of power) can do is assist, support and mentor other women. Women who have "made it" into senior positions can play a critical role in fostering female-friendly workplace environments and in enabling other women to participate in positions of authority. They can do this through concrete measures such as establishing diversity-sensitive workplace policies, nurturing respectful workplace cultures and not tolerating workplace abuses. Encouragingly, many women I spoke to who felt that they had managed to surmount the difficulties of succeeding as a woman in an IT-related field also felt that they had an obligation to help other women gain the skills and enjoy the opportunities available to them. Tatiana, a graduate student who was doing research into girls and technology, saw it as "her duty" to help out girls and other women through volunteer instruction. "I am very capable to work in IT fields, but I was never encouraged to work in them. I see it as my duty to try and help someone else ... Nothing compares to seeing a girl so excited and working off the computer, and being successful at whatever project she was trying to do." Lilith tells me that she keeps an eye out for talented women to recruit for her company and to whom she often offers free skills training and mentorship. "I give other women the respect I feel they deserve," she says. "That alone puts me way ahead of the guys."

However, this support should not be limited to individual workers. It is hard to build a strong, diverse IT work force when IT jobs are precarious, unstable, short-term and poorly paid, and when workers are divided against one another along fault lines of social and geographic location, as in the case of regional outsourcing, where white male North American programmers are pitted against Indian programmers who are often female. In fact, *Wired* magazine ran a feature explicitly comparing the two groups, which read like the sportscast of a grudge match between the "pissed off programmer" and the Indian female engineer who will "do your $70,000-a-year job for the wages of a Taco Bell counter jockey."[12] "Kiss your cubicle good-bye," *Wired*'s cover intones ominously. Non-white and female workers used to be simply absent from IT cultural representation. Now, just as they were a century ago, they're often seen as the job-stealing enemies who are willing to work for less money and undercut white male professionals. The substantive global inequalities of IT work (including the low-cost labour that manufactures the gadgets that spur technological consumerism) and the importance of sustainable, decent livelihoods for all workers remain unaddressed. Anger and critical attention is shifted away from IT corporations concerned with executive self-preservation and short-term profit and away from the global inequalities that make poor regions so appealing to North American corporations.

IT employers and institutions who are concerned with improving women's role in IT have a responsibility to ensure that the material conditions are in place to make it happen. This includes looking at the IT workforce using a diversity lens. Where are there absences in the technical team? Who is doing what and how are they doing it? What skills are being valued, and is this an accurate, fair perception? What structures need to be in place so that diversity can occur? Is short-term profit being privileged over the long-term gains that can occur with a vibrant, heterogeneous, well-compensated workforce? More and more women, especially younger women, are choosing IT as a career field. Some women are also moving into IT from other occupations. But numbers are not enough. Women, and men, need to demand *quality* of work as well as *quantity.* We cannot just push women into IT if we want them to be there. We have to pull them in with decent work, a welcoming workplace and improved choices. We also need to change the cultural elements of IT work that encourage long hours, fail to provide strong worker protections, sustain an imbalance between work and life and maintain hierarchies based on social and geographical location as well as assumed technical competence. Making IT a more hospitable place for those people who have traditionally been marginalized in IT will result in a more positive environment for all workers.

Appendix A

Data on Paid and Unpaid Work

Section A.1: IT Industry Data

Table 1 Top Five IT-Related Industries for Women in Canada, 2003

Type	Industry	Number of workers (thousands)
Services	Telecommunications	54
Services	Computer systems design & related services	53.4
Manufacturing	Semiconductor & other electronic components	16.6
Services	Computer & communications equipment wholesalers/distributors	11.7
Manufacturing	Computer & peripheral equipment	10.2

Source: Statistics Canada, *Labour Force Survey*, custom tabulation, 2003

Table 2 Shares of Men and Women in IT-Related Industries, 2003

Note: this table orders industries from highest to lowest percentage of women

Industry	Men	Women
Computer and electronic product manufacturing		
Semiconductor & Other Electronic Component Manufacturing	54.3%	45.7%
Audio & Video Equipment Manufacturing	57.1%	42.9%
Computer & Peripheral Equipment Manufacturing	59.8%	39.8%
Communications Equipment Manufacturing	66.2%	33.8%
Commercial & Service Industry Machinery Manufacturing	68.8%	31.2%
Other Electrical Equipment & Component Manufacturing	69.2%	30.1%
Navigational, Medical & Control Instruments Manufacturing	75.9%	23.5%
Information and communications technologies services		
Data Processing Services	37.9%	62.1%
Telecommunications	56.1%	43.9%
Software Publishers	62.4%	36.6%
Computer & Communication Equipment & Supplies Wholesalers/Distributors	66.4%	33.6%
Computer Systems Design & Related Services	70.3%	29.7%
Electronic & Precision Equipment Repair & Maintenance	82.4%	17.6%

Source: Statistics Canada, *Labour Force Survey*, custom tabulation, 2003

Section A.2: IT Occupation Data

Table 3 Shares of Men and Women in IT-Related Occupations, 2001

Note: this table orders occupations from highest to lowest percentage of women

Occupation	Men	Women
Data entry clerks	18.40%	81.60%
Desktop Publishing Operators and Related Occupations	33.50%	66.50%
Electronics Assemblers, Fabricators, Inspectors and Testers	46.10%	54.00%
Database Analysts and Data Administrators	57.70%	42.30%
Systems Testing Technicians	60.00%	40.10%
Supervisors, Electronics Manufacturing	62.90%	37.10%
Web Designers and Developers	66.70%	33.30%
Information Systems Analysts and Consultants	68.60%	31.40%
User Support Technicians	68.90%	31.10%
Computer and Information Systems Managers	73.50%	26.50%
Computer and Network Operators and Web Technicians	74.60%	25.50%
Computer Programmers and Interactive Media Developers	76.50%	23.50%
Software Engineers	81.60%	18.40%
Computer Engineers (Except Software Engineers)	85.20%	14.90%
Engineering Managers	89.10%	10.90%

Source: Statistics Canada, *Census of Population 2001*

Table 4 Average Annual Earnings in IT-Related Occupations by Gender, and Average Wage Gap, 2001

Note 1: This table orders occupations by the amount of the wage gap, from highest to lowest.
Note 2: Earnings includes both salary for employees and income for self-employed workers.
Note 3: Wage gap is calculated as the percentage of women's earnings relative to men's.

	Average annual salary or self-employment income		
	Men	Women	Wage gap
Engineering managers	$84,657	$52,474	62.0%
All other non-IT occupations	$38,610	$24,586	63.7%
All IT Occupations subtotal	$49,707	$34,554	69.5%
Supervisors, electronics manufacturing	$43,862	$30,777	70.2%
Desktop publishing operators and related occupations	$31,398	$23,725	75.6%
Computer engineers (except software engineers)	$62,305	$47,120	75.6%
Database analysts and data administrators	$50,728	$39,213	77.3%
Software engineers	$62,789	$49,712	79.2%
Computer and information systems managers	$76,316	$61,504	80.6%
Computer and network operators and web technicians	$40,019	$34,038	85.1%
Electronics assemblers, fabricators, inspectors and testers	$25,479	$21,947	86.1%
Information systems analysts and consultants	$55,740	$49,228	88.3%
Computer programmers and interactive media developers	$44,419	$39,654	89.3%
User support technicians	$36,376	$33,003	90.7%
Web designers and developers	$29,601	$29,078	98.2%
Systems testing technicians	$38,277	$38,716	101.1%
Data entry clerks	$19,163	$21,146	110.3%

Source: Statistics Canada, *Census of Population 2001*

Chart 1

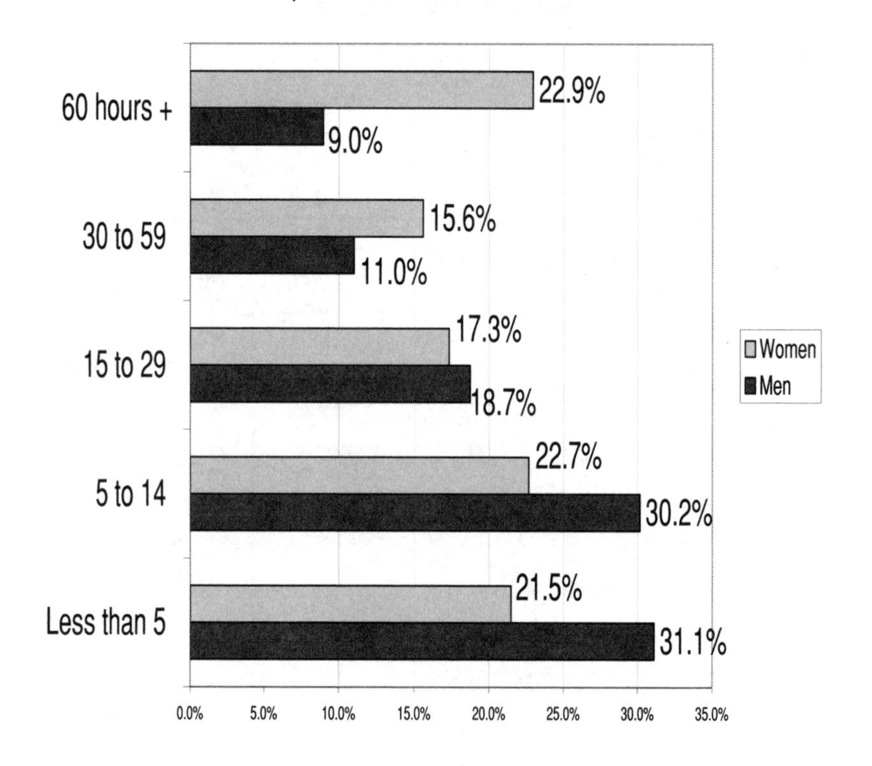

Source: Statistics Canada, *Census of Population 2001*

Chart 2

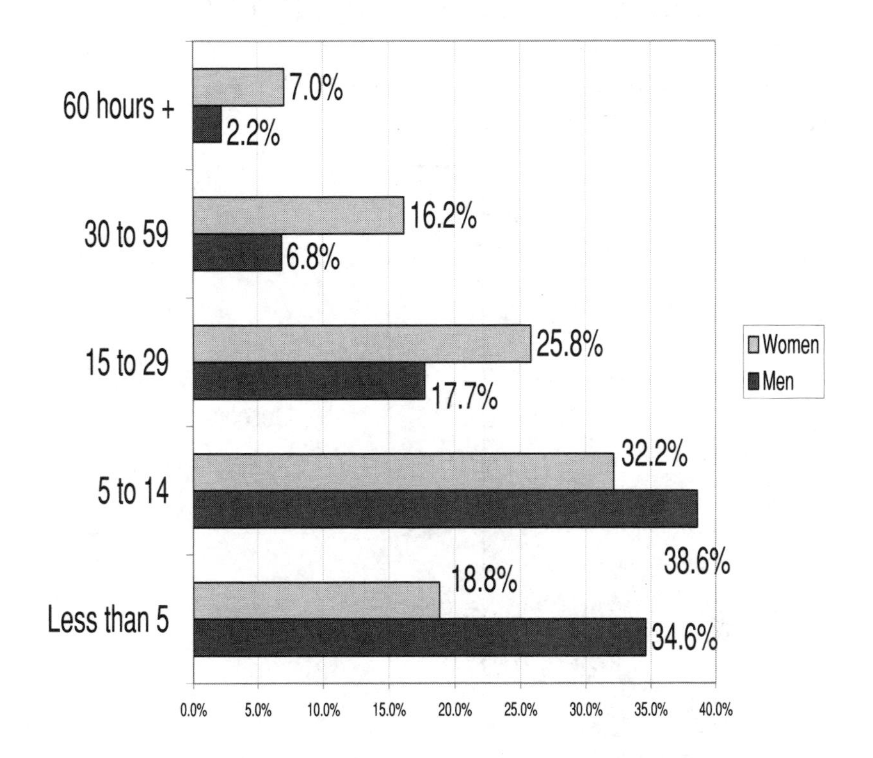

Source: Statistics Canada, *Census of Population 2001*

Chart 3

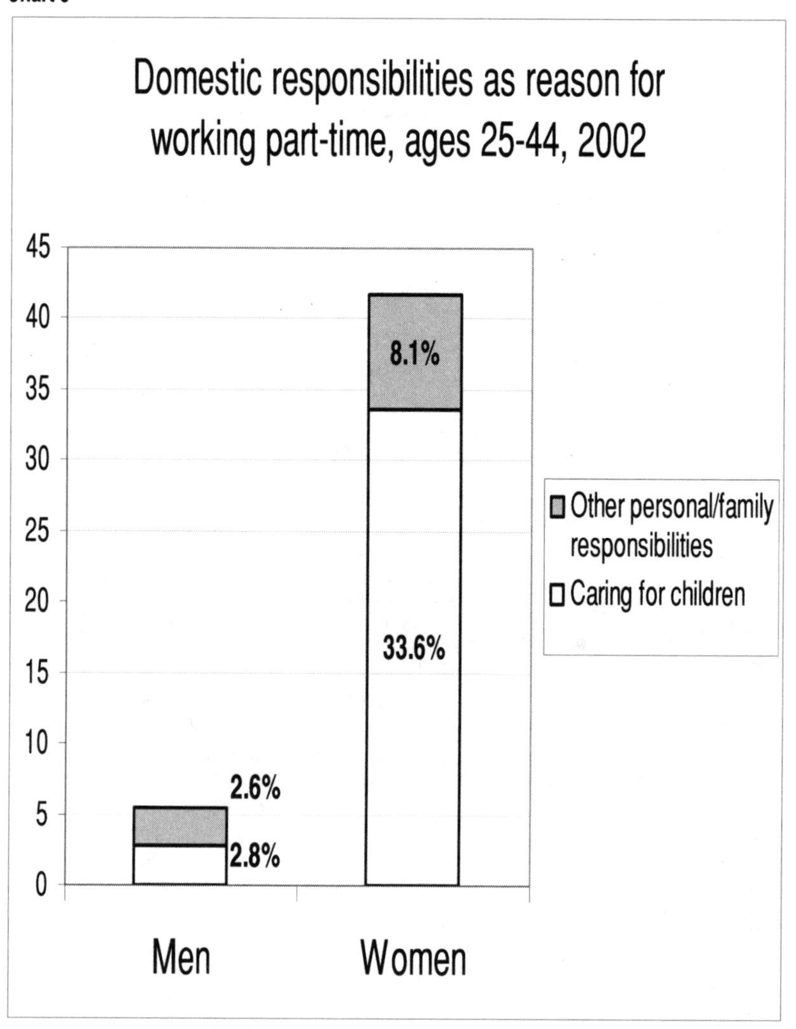

Source: Statistics Canada, *Labour Force Survey*, custom tabulation, 2002

APPENDIX B

INTERVIEW PARTICIPANTS

Name given in text: With the exception of Tari and Laurie, who consented to have their real names published, all other names are pseudonyms.

Form of employment: FT = full-time, PT = part-time, SE = self-employed

N/S refers to not stated: This means that either the interview subject did not want to respond, or that for some reason, the question was not asked.

Marital status: CL = common law, M = married, S = single, D = divorced

Name given in text	*Age*	*Marital status*	*Children*	*Education*	*Job title*
Ann	26	CL	0	BA English	Marketing account manager
	32	M	0	BA Radio and TV	Project managerr
	25	S	0		Client services rep andWeb design
Brenda		S	1	High school, some college	Admin Asst, E-Business developer
Charity	27	CL	0	BA New Media	Content co-ordinator
	32	S	0	Partial BA	Communications strategist
Karen	26	M	0	High school	Assistant Web designer/Webmaster
Janice	42	S	1	Some university	Process documentation specialist
Nadine	40s	S	0	College diploma	Trainer
Katherine	N/S	M	2		
Carmen	41	CL	0	MA Information Studies	Information architect
	35	M	0	2 years science undergrad, BA English/Humanities	Manager, Internet initiatives

	N/S	N/S	0	BA Sociology and Art History	CEO of own company
Sandra	32	M	0	Partial BA	Senior consultant
Maggie	42	S	0	BA Arts, College certificate in Radio/TV	Technical journalist
	31	CL	0	BA Journalism, some college	Acting Director, Media Relations
	36	S	0		
Melanie	31	S	0	BA	VP of client-partner relations
Donna	60	M	0		
	36	M	0	BSc	Writer
	36	M	0	BA Humanities/Film	Sales administrator
	N/S	N/S	0		
Lilith	29	S	0	High school, some university	Manager, network engineering
Mary	31	M	1	BA Journalism	Freelance writer/editor
Donna	30	N/S	0		
Gillian	30	S	0	BA Psychology	Project manager
	46	N/S	0		CD-ROM producer and developer
Erin	29	CL	0	BSc, MA Environmental Science	Resource system intranet facilitator
	26	S	0	Student, college	Junior informatics specialist

APPENDIX B

Preeti	32	M	0	BA Philosophy	Unemployed, seeking self-employment
Helen	41	M	2	College diploma, Fine Arts	Graphic designer
	N/S	M	0		
Cynthia	32	S	0	Various industry certifications in accounting and networking	Customer support manager
Susan	58	S	2	BA Arts	Communications consultant
	30	M	0	BA Political Science, MA Information Studies	Information architecture specialist
Sarah	29	CL	0	Sheridan, textiles, cpu graphics	Graphic designer
	30	CL	0	BSc Psychology and Sociology	Medical writer
Patricia	49	D	2	Grade 9	Project developer
Neetha	44	M	2	BA Math/Stats, M. Econ, pending CGA	Accounting/funding manager
	27	M	0	BA Psychology	IT consultant
	26	M	0	BA English/Publishing	Project manager - Production
	39	M	2		Multimedia Specialist
Tatiana	29	M	1	MA OISE, Social/Equity Stud in Edn	Graduate student
Anya	30	S	0	BSc Biochemistry	Web developer/Flash developer
	31	M	0		Web designer
Debbie	25	CL	0	Some university	Intranet developer

Irene	41	D	2	High school, 1 yr college	Administrative support
Jamie	33	CL	0	BA English/Cultural Studies, BA Radio/TV	Technical writer
Jane	38	S	0	BA Business	E-Commerce manager
	54	D	1	BA Psychology, College diplomas in English/Media, Management Development	
	31	CL	0		Senior software consultant
Farideh	39	S	0	MA	Graduate student
Lina	28	M	0	MA IT management	Unemployed, seeking employment
Tari	40	S	0	BA Business Administration	Call-centre worker, self-employed Web developer
Mike	30	S	0	BFA	Call-centre worker, former techworker in Y2K
Maria	53	M	3	Certificate, Publishing	Content and information design manager
Laurie	48	S	1	College diploma, electronics engineering, pending BA in IT	Student
Rebecca	36	CL	0	BA Art History and Museum Studies	Recently laid off, formerly program manager

Naomi	32	S	0	BA, working on IT certification	Student
Faith	39	M	0	BA Communications, MA Sociology	Marketing manager
Diane	28	CL	0	BA Geography	Web designer
Angie	20	S	0	Working on BA	Student

Notes

Introduction

1. IT is sometimes referred to as the information and communications technology industry, or ICT industry. IT in this text is also used to refer to "new media." "New media" tends to be used to refer to things like on-line advertising, interactive kiosks, multimedia (something incorporating still graphics, video, hyperlinks, etc.), on-line communities, software development, content creation, design, management, promotion, distribution and electronic commerce (or "e-commerce"). A recent Statistics Canada publication defines the ICT sector as "the combination of manufacturing and services industries, which electronically capture, transmit, and display data and information." Heidi Ertl, *Beyond the Information Highway: Networked Canada* (Ottawa: Statistics Canada, Science, Innovation, and Electronic Information Division, and Minister of Industry, 2001), 12.

2. Ibid., 14.

3. "Friction-free economy" is a term coined by Theodore Lewis in his book *The Friction-Free Economy: Marketing Strategies for a Wired World* (New York: HarperBusiness, 1997). In the Preface, he writes, "It may be too fast for royalty, but the software economy is on its way. It may be a mystery to the establishment, but it is well understood by the Netheads2 in Wired World3. It may violate the doctrine handed down by classical economists, but it does follow a set of laws. It may be just in time." The quote "where demand effortlessly follows production" comes from an on-line article by Lewis entitled, "Alice in Wired World 7: The Positive Feedback of the Software Economy." Available on-line from www.friction-free-economy.com/archives/18-7.html.

4. For an excellent discussion of the implications of this notion, see Evelyn M. Hammonds, "New Technologies of Race," in Jennifer Terry and Melody Calvert, eds., *Processed Lives: Gender and Technology in Everyday Life* (New York: Routledge, 1997), 107–122; and Lisa Nakamura, "Where Do You Want to Go Today? Cybernetic Tourism, the Internet, and Transnationality," in Beth Kolko, Lisa Nakamura and Gilbert B. Rodman, eds., *Race in Cyberspace* (New York: Routledge, 2000), 15–26.

5. Advertisement for Aspira (Motorola), appearing in *Wired,* January 2000, 15.

6. Trevor Haywood, "Global Networks and the Myth of Equality: Trickle Down or Trickle Away?" in Brian Loader, ed., *Cyberspace Divide: Equality, Agency, and Policy in the Information Society* (New York: Routledge, 1998), 19–34. Haywood focuses on the development of technological networks and the lack of political debate about their manifestations and effects, as well as the problem of access and power relations. See also Mike Holderness, "Who Are the World's Information-Poor?" in Loader, ed., *Cyberspace Divide,* 35–56.

7. See, for example, Ursula Franklin, *The Real World of Technology* (Toronto: House of Anansi Press, 1999). See also David Noble, *Progress without People: New Technology, Unemployment, and the Message of Resistance* (Toronto: Between the Lines Press, 1995).

8. Anne Balsamo, "Public Pregnancies and Cultural Narratives of Surveillance," in *Technologies of the Gendered Body: Reading Cyborg Women* (Durham: Duke University Press, 1996), 80–115.

9. Geoff Bowlby and Stephanie Langlois, "High-Tech Boom and Bust," *Perspectives on Labour and Income* (2002), 12–18.

10. For example, formerly thriving telecommunications companies are suffering. At the end of May 2002 in the United States, Moody's downgraded AT&T stock to two notches above junk, which brought it down to the level of Sprint. All leading U.S. rating agencies cut WorldCom's status to junk bonds (and later WorldCom filed for U.S. Chapter 11 bankruptcy protection), and Standard and Poor dropped Qwest to junk. Jonathan Stempel, "AT&T Cut to Two Notches Above Junk," *Reuter's Business Report* (on-line edition), 29 May 2002. In Canada, the most familiar company downturn was Nortel, which cut approximately half its workforce (50,000 people) in just over a year, and whose stock has lost about 90 percent of its highest value. See Associated Press (on-line edition), "Nortel To Cut Another 3,500 Jobs," 29 May 2002; and Susan Taylor, "Nortel Deepens Its Cuts Amid Technology Slump," *Reuters Technology Report* (on-line edition), 29 May 2002.

11. Bowlby and Langlois report that approximately 14,000 men's jobs were lost while 20,000 women's jobs were lost. Given women's major under-representation in the field, this is very significant.

12. As I write this, MCI, the second biggest long-distance telephone carrier in the United States, and formerly known as WorldCom, is under investigation for fraud. U.S. Justice Department officials are examining evidence that "MCI may have in effect 'laundered' calls through small phone companies and redirected some domestic calls through Canada to avoid paying access fees or to shift them to rival long-distance carriers, according to people involved in the investigation." Geoff Gibbs, "WorldCom Inquiry Haunts MCI," *The Guardian* (on-line edition), 28 July 2003.

13. In the early 1600s in Europe, a mania for tulip bulbs emerged after the plants were introduced from Turkey. Estimates of the price of a single valuable tulip

bulb ranged from $10,000 to $75,000 in our dollars. In the late 1630s, facilitated in part by the finance regulations then in place in Amsterdam, the bottom fell out of the tulip bulb market, and many investors who had gambled all of their assets on bulb speculation went bankrupt. The connection between tulip-bulb and tech-stock investing has been noted by several sources. For example, the CIC Group, an international finance services company, even put out a document in 2000 called *Tulips and Tech Stocks* (March 13, 2000). This document urges informed caution among would-be tech investors.

14. The archive-minded technophile can even obtain old stock certificates from bankrupt or downgraded dot-coms from www.scripophily.net. Up for grabs include Amazon.com and Bingo.com.

15. Such as WashTech in Washington and Alliance@IBM for IBM workers.

16. Tahany Gadalla, "Are More Women Studying Computer Science?" *Resources for Feminist Research/Documentation sur la recherche féministe* 27, no. 1/2 (1999), 137–142; Jane Margolis and Allan Fisher, *Unlocking the Clubhouse: Women in Computing* (Cambridge, MA: MIT Press, 2002).

17. Human Resources Development Canada, *Achieving Excellence: Investing in People, Knowledge and Opportunity* (Ottawa: Industry Canada, 2001), 6, 14. Emphasis in original.

18. "The call centre is … characterised by the integration of telephone and VDU [video display unit] technologies." Phil Taylor and Peter Bain, "'An Assembly Line in the Head': Work and Employee Relations in the Call Centre," *Industrial Relations Journal* 30, no. 2 (1999), 102.

19. Ruth Buchanan and Sarah Koch-Schulte, *Gender on the Line: Technology, Restructuring, and the Reorganization of Work in the Call Centre Industry* (Ottawa: Status of Women Canada, 2001); Ruth Buchanan and Joan McFarland, "The Political Economy of New Brunswick's 'Call Centre' Industry," *Socialist Studies Bulletin* (1997), 1–40.

20. For example, women tend to put in less overtime than men. See Information Technology Association of Canada (ITAC), *Meeting the Skills Needs of Ontario's Technology Sector: An Analysis of the Demand and Supply of IT Professional Skills* (Toronto: IDC Canada and Aon Consulting, 2002).

21. The ITAC report puts their numbers at 24 percent of the IT industry workforce. Statistics Canada, using 2001 Census data, puts the number of women in IT at 27 percent. Roman Habtu, "Information Technology Workers," *Perspectives on Labour and Income* (July 2003), 5.

22. ITAC report indicates that 23 percent of women working in IT said their choice was "accidental," compared with 14 percent of men. Women in my interviews confirmed this perspective as a common one.

23. Cynthia Cranford, Leah Vosko and Nancy Zukewich, "The Gender of Precariousness in the Canadian Labour Force," *Industrial Relations/Relations Industrielles* (Fall 2003), 454–482.

24. Heather Dryburgh, *Changing Our Ways: Why and How Canadians Use the Internet* (Ottawa: Statistics Canada, 2000), 12.

25. Mindy Gewirtz and Ann Lindsey, *Women in the New Economy: Insights and Realities* (Brookline, MA: GLS Consulting, 2000), 2.

26. Boel Berner and Ulf Mellström, "Looking for Mister Engineer: Understanding Masculinity and Technology at Two Fin De Siécles," in Boel Berner, ed., *Gendered Practices: Feminist Studies of Technology and Society* (Linkoping, Sweden: Dept. of Technology and Social Change, 1997), 39–68; Tove Håpnes and Knut Sørensen, "Competition and Collaboration in Male Shaping of Computing: A Study of Norwegian Hacker Culture," in Keith Grint and Rosalind Gill, eds., *The Gender-Technology Relation: Contemporary Theory and Research* (London: Taylor and Francis, 1995), 174–191.

27. I interviewed sixty-two women during the research and writing of my PhD dissertation "From Web Grrls to Digital Eve: The Gendered Practice of Information Technology Work and Organization" (York University, 2001). Their quotes are used here with their permission. For a breakdown of who the women are in this book, see Appendix B.

CHAPTER 1: WOMEN'S IT WORK IN CONTEXT

1. Eugenia Date-Bah, "Introduction," in Eugenia Date-Bah, ed., *Promoting Gender Equality at Work: Turning Vision into Reality in the Twenty-First Century* (New York: Zed Books, 1997), 5. Pat Armstrong and Hugh Armstrong concur: "The increasing visibility of women outside the home and the emphasis on female attainment of jobs at the top of the career ladder have camouflaged the lack of basic change in most women's work." Pat Armstrong and Hugh Armstrong, *The Double Ghetto: Canadian Women and Their Segregated Work,* 3d ed. (Toronto: McClelland and Stewart, 2001), 13.

2. Marjory MacMurchy wrote optimistically in 1919, "It is evident that women have always worked, and worked hard ... The increasing opportunities of girls ... in paid employment are likely to become a contributing factor in the humanizing of every form of industry." Marjory MacMurchy, *The Canadian Girl at Work: A Book of Vocational Guidance* (Toronto: A.T. Wilgress, 1919), iv, vi. Dionne Brand remarks somewhat less enthusiastically about Black women's paid labour in Canada, "Simply put, Black women needed the work — they needed the money — and waged work had [historically] been an essential part of their daily lives." Dionne Brand, "'We Weren't Allowed to Go Into Factory Work until Hitler Started the War': The 1920s to the 1940s," in Peggy Bristow, ed., *We're Rooted Here and They Can't Pull Us Up: Essays in African Canadian Women's History* (Toronto: University of Toronto Press, 1994), 181.

3. See Appendix A at the back of this book for data from *Labour Force Survey* 2002 and the 2001 Census.

4. However, one methodological difficulty of quantifying women's domestic labour is the practice of data collection, which, according to Date-Bah, tend to "undercount women's economic activities and to under-assess their economic contribution, especially their non-market and informal work." Date-Bah, "Introduction," *Promoting Gender Equality at Work, 2.*

5. Hestia, in Greek mythology, was a goddess associated with homemaking and household management. Her role of keeping the home fires lit signifies this domestic role.

6. See Chart 3 in Appendix A at the back of this book.

7. Bruce Arai, "Self-Employment as a Response to the Double Day for Women and Men in Canada," *Canadian Review of Sociology and Anthropology* 37, no. 2 (May 2000), 137.

8. Ibid., 139.

9. The first incarnation of the Commodore Amiga, the A1000, appeared in 1985.

10. Betty Friedan, *The Feminine Mystique* (New York: Dell, 1963), 229.

11. Ibid., 236.

12. Critics have since noted, correctly, that Friedan omitted any discussion of women for whom the home was both workplace and sanctuary. Black women, for example, had few options for employment besides domestic service, but their own home could be a respite from the racism they experienced in public. It is not clear here whether or not Friedan is subtly critiquing the devaluation of "women's work" in the labour force or merely using the ideology of women's paid work as "help" to point out the absence of professional (read higher-status) careers for women. In any case, the existence of women whose experience did not fit this model, such as women who worked full-time for wages their entire life, was erased. Moreover, the economic primacy of the male breadwinner was not questioned.

13. Virginia Woolf, *A Room of One's Own* (1929; reprint, London: Grafton, 1977), 29.

14. Ibid., 58.

15. Advertisement for Hitachi appearing in *Wired,* February 1997, 15.

16. Advertisement for Philips Magnavox appearing in *Wired,* February 1997, 134.

17. Advertisement for Kasper clothing, appearing in *Vogue,* August 2003, 39.

18. Friedan, *The Feminine Mystique,* 11.

19. Statistics Canada, *2001 Census.* The 2001 *Survey of Labour and Income Dynamics* (SLID) indicates that 67.7 percent of Canadian women workers (not including the self-employed) have children younger than eighteen. For

the self-employed, the *Survey of Self-Employment* shows that 55.3 percent of self-employed women have children younger than twenty-four. The *2001 Census* indicates that 60.5 percent of all Canadian women who are part of family households have children living at home. This includes women who are married, common-law or lone parents.

20. Virginia Galt, "Is There a Home Front Advantage?" *The Globe and Mail,* 16 May 2003, C1, reporting on the report *Leaders in a Global Economy: A Study of Executive Women and Men,* co-authored by Boston College and Catalyst. Seventy-five percent of male executives had stay-at-home spouses, while 74 percent of the women had spouses who worked full-time.

21. Bonnie Anderson and Judith Zinsser, *A History of Their Own* (New York: Oxford, 2001).

22. For an interesting discussion of the development of waged industrial work in Ontario through the late nineteenth and early twentieth centuries, see Joy Parr, *The Gender of Breadwinners: Women, Men and Change in Two Industrial Towns, 1880-1950* (Toronto: University of Toronto Press, 1990). Also see Dionne Brand, *No Burden to Carry: Narratives of Black Working Women in Ontario, 1920s–1950s* (Toronto: Women's Press, 1991).

23. As Germaine Greer noted, "More than half the housewives in this country work outside the home as well as inside it because their husbands do not earn enough money to support them and their children at a decent living standard ... And yet the myth is not invalidated as a myth ... The women of the lower classes have always laboured, whether as servants, factory hands or seamstresses or the servants of their own households, and we might expect that the middle-class myth did not prevail as strongly in their minds. But it is a sad fact that most working-class families are following a pattern of 'progress' and 'self-improvement' into the ranks of the middle class ... They too consider even if they cannot exactly manage it that mum *ought* to be at home keeping it nice for dad and the kids." Germaine Greer, *The Female Eunuch* (London: Paladin, 1970), 216.

24. Alvin Toffler, *The Third Wave* (New York: Morrow, 1980).

25. Heather Menzies, "Telework, Shadow Work: The Privatization of Work in the New Digital Economy," *Studies in Political Economy* 53 (Summer 1997), 103–125.

26. Heather Dryburgh, "Immigrant Science and Technology Workers: Employing the Brain Gain?" paper presented at the Congress of Social Sciences and Humanities, June 2003, Halifax, NS. See also Roman Habtu, "Information Technology Workers," *Perspectives on Labour and Income* (2003), 5–11.

27. Examining 1999 Statistics Canada data, we can see that the largest employer of women is health care and social assistance (81 percent of all workers in that industry and 17.5 percent of all women workers), followed by education (63.8 percent of all workers and 9.4 percent of all women workers), finance, insurance, real estate (60 percent of all workers and 7.7 percent of all women

workers), trade (47.8 percent of all workers, and 16 percent of women workers) and public administration (46.4 percent of all workers and 5 percent of all women workers). Statistics Canada, "Employment by detailed industry and sex, 1999," *Labour Force Historical Review, 2000.* In contrast, women represent only 18 percent of workers in natural and applied sciences and related occupations. Statistics Canada, 1996 Census.

28. Armstrong and Armstrong, *The Double Ghetto.*

29. See also Marilyn Waring, *Counting for Nothing: What Men Value and What Women Are Worth* (Toronto: University of Toronto Press, 1999) for an expansion on this theme.

30. Dryburgh, "Immigrant Science and Technology Workers," 8. Italics in original.

31. Habtu, "Information Technology Workers," 9. Statistics Canada, *Census 2001.*

32. Ibid., 6.

33. Ibid., 8. The other areas with high concentrations of IT workers are Ottawa-Gatineau, Montreal, Vancouver and Calgary.

34. Judy Fudge, "Flexibility and Feminization: The New Ontario Employment Standards Act," *Journal of Law and Social Policy* 16 (2001), 1–22.

35. Canadian Labour Congress, "The Widening Wage Gap Between Women and Men," *Canadian Labour Congress Economy* 14, no.1 (Spring 2003), 10. This article uses data from Statistics Canada's *Labour Force Survey* and covers full-time, full-year workers.

36. Statistics Canada, *Census 2001;* Armstrong and Armstrong, *The Double Ghetto.* The Canadian Labour Congress reports that the February 2001 ICFTU's "Ask a Working Woman Survey" indicated that the number one concern of all working women was higher pay. Coming in a close second was the desire for more control over work hours.

37. Some research indicates that wages for immigrant women have increased during the 1990s, but that in 1998 they still made, on average, $3.20 per hour less than Canadian women, and $7.20 less than non-immigrant men. Their average earnings were "61% of the wages and salaries of other women, and little more than one-third of the average annual earnings of non-immigrant men." Ekuwa Smith and Andrew Jackson, *Does a Rising Tide Lift All Boats? The Labour Market Experiences and Incomes of Recent Immigrants* (Ottawa: Canadian Council on Social Development, 2002), 10–11. However, other research indicates that, in fact, immigrant women's wages declined. See John Shields, "No Safe Haven: Markets, Welfare, and Migrants," paper presented to the Canadian Sociology and Anthropology Association, Congress of the Social Sciences and Humanities, Toronto, June 2002; Statistics Canada, "The Changing Profile of Canada's Labour Force," *2001 Census: Analysis Series* (Ottawa, 2003).

38. Marie Drolet, *Wives, Mothers, and Wages: Does Timing Matter?* (Ottawa: Statistics Canada, Business and Labour Market Analysis Division, 2002).

39. Marie Drolet, *The Persistent Gap: New Evidence on the Canadian Gender Wage Gap* (Ottawa: Statistics Canada, Income Statistics Division, 2001), and *The Persistent Gap: Evidence on the Canadian Gender Wage Gap* (Ottawa: Statistics Canada, Income Statistics Division, 1999). Emphasis added.

40. See Appendix A at the back of this book for exact figures on paid and unpaid work.

41. Wage gap citations for average wage gap, computer and software engineers, engineering managers, systems analysts and computer programmers from Statistics Canada, *Census 2001.* Also see Lorraine Dyke, Linda Duxbury and Natalie Lam, *Gender Differences in High Tech Careers,* report prepared for the Gender Analysis and Policy Directorate (Ottawa: Human Resources and Development Canada, 2001). This report shows that women IT workers make approximately 72 to 79 percent of men's wages.

42. Habtu, "Information Technology Workers," 8. Habtu puts the median annual earnings for Web designers at $29,100. Publicly available 2001 Census data do not provide medians but set the average wage for female Web designers at $28,815.

43. For example, John Anderson writes: "Technology has been generally used as a weapon, not for our liberation from monotony, stress and want, but rather for the private appropriation of profit; for changing the workplace beyond recognition with the main aim of increased revenues instead of for the greater welfare of society and better conditions for working people." John Anderson, "Technology on Trial: Lessons from the Labour Frontiers," in *Re-Shaping Work: Union Responses to Technological Change* (Don Mills, ON: Ontario Federation of Labour, 1995), 9.

44. Ursula Franklin, *The Real World of Technology* (Toronto: House of Anansi Press, 1999).

45. Ibid.

46. See, for example, Harry Braverman's work which, while useful, is limited in its application to the present context. *Labour and Monopoly Capital: The Degradation of Work in the Twentieth Century* (New York: Monthly Review Press, 1974).

47. Sally Hacker's research provides a useful point of departure for discussing shifts in the organization of work with technology, and the importance of a perspective which is attentive to structural relations of gender, race, class and other factors. In her germinal 1970s research on women's work in telecommunications (one of the precursors to current IT work), she focused on gender stratification as applied to technological displacement of work. She argued that "the impact of technological change on women's work varies by the political and economic framework of particular technologies, by the stage of development within such frameworks, and by linkages between

family and work" (45). Studying job shifts at AT&T, she concluded that women's positions were largely displaced as a result of work reorganization, and that affirmative action backlash for white men was a myth. As women moved into male-dominated arenas, such as equipment maintenance and installation, those arenas were deliberately phased out. While some small gains for women of colour were made in clerical at time of the study writing, in general, noted Hacker, the intersection of gender and race resulted in "displacement [which] struck most sharply where minority women worked" (58). Women's attempts at organization were thwarted not only by hostile management but also by male workers, husbands and union officials. Most significantly for the purposes of this section, Hacker showed how senior staff at AT&T intentionally used divisions of sex and race among workers to enhance corporate profit and create transitions to a technological base. Her eventual conclusion was that with some variations, "the flow of workers through occupations during technological change appeared to proceed from white male to minority male to female, then to machines" (111). Sally Hacker, "Sex Stratification, Technology, and Organizational Change: A Longitudinal Case Study AT&T," in Dorothy Smith and Susan Turner, eds., *Doing It the Hard Way: Investigations of Gender and Technology* (London: Unwin Hyman, 1990), 45–67.

Chapter 2: The Struggle for Skills

1. The "points" test for skilled workers looks at several criteria: education, language proficiency, experience, age, potential for employment in Canada and adaptability to Canadian life.

2. MCSE is a Microsoft Certified Systems Engineer. CCNA is Cisco Certified Network Associate. However, in keeping with the subversive nature of IT culture, many people in the industry joke that MCSE stands for things like Must Consult Someone Experienced or Making Computers Slow Everyday.

3. Human Resources Development Canada and Industry Canada, *Achieving Excellence: Investing in People, Knowledge, and Opportunity* (Ottawa: Industry Canada, 2001), 6.

4. Information Technology Association of Canada (ITAC), *Meeting the Skills Needs of Ontario's Technology Sector: An Analysis of the Demand and Supply of IT Professional Skills — Executive Summary* (Ottawa: IDC Canada and Aon Consulting, 2002), 9.

5. Ibid., 11.

6. Garnett Picot and Andrew Heisz, *The Performance of the 1990s Canadian Labour Market* (Ottawa: Statistics Canada, Business and Labour Market Analysis Division, 2000), 21.

7. Canadian Advisory Council on Science and Technology, *Stepping Up: Skills and Opportunities in the Knowledge Economy* (Ottawa: Industry Canada, 2000), 2.

8. Ibid., 31.

9. Vladimir Lopez-Bassols, *ICT Skills and Employment* (Paris: Organization for Economic Cooperation and Development, 2002), 12. The OECD is a policy and research organization of thirty countries which is dedicated to "democratic government and the market economy."

10. Jane Margolis and Allan Fisher, *Unlocking the Clubhouse: Women in Computing* (Cambridge, MA: MIT Press, 2002), 65.

11. A job search Web site for IT workers is called FreeAgent.com.

12. As found originally in Alfred Marshall's 1920 work *Principles of Economics* (1920; reprint, London: MacMillan, 1961). For a more extensive discussion of this, see Michele Pujol, *Feminism and Anti-Feminism in Early Economic Thought* (Vermont: Edward Elgar Publishing, 1992).

13. Interestingly, the role of women in the original version of the human capital theory, according to Marshall, was to sacrifice paid employment for nurturing their children's (one presumes their sons') human capital. Thus, women's investment in their human capital has the purpose of engendering the development of human capital in their children. It is reminiscent of Mary Wollstonecraft's argument about the purpose of education for women in *A Vindication of the Rights of Woman* (1792): to create better mothers who will rear better (again presumably male) citizens.

14. See, for example, Dionne Brand, "We Weren't Allowed to Go Into Factory Work until Hitler Started the War': The 1920s to the 1940s," in Peggy Bristow, ed., *We're Rooted Here and They Can't Pull Us Up: Essays in African Canadian Women's History* (Toronto: University of Toronto Press, 1994), 171–191. Alfred Marshall also called for women's minimum wage to be lower than men's as a way to discourage them from working outside the home. The immediate sacrifice to women's economic health is thought to have dividends in the economic health of future generations who have their human capital nurtured by full-time mothers.

15. For example, women of colour who are recent immigrants to Canada are preferentially recruited for difficult, low-status, low-waged jobs such as garment work, domestic work and agricultural labour. Heather MacIvor, *Women and Politics in Canada* (Peterborough, ON: Broadview Press, 1996), 117. The social construction of a category called "immigrant women" has particular implications for these women's work choices, in that it "facilitates the construction of ethnic segments in the labour market where knowledge of English, Canadian education and work experience emerge as structural barriers to move on to better paying, higher status jobs." See Ayse Nur Oncu, "Informal Economy Participation of Immigrant Women in Canada" (MA thesis, University of Alberta, 1992), 26.

16. As Oncu notes, "The need for cheap sources of labour by employers in the secondary segment continuously perpetuates the ethnic, racist, and sexual stereotyping in the society." Ibid., 26.

17. See, for example, Roxana Ng's work cited in the Bibliography and "Immigrant Women: The Construction of a Labour Market Category," *Canadian Journal of Women and the Law* 4 (1990), 96–113.

18. Thirty-six percent of all immigrant men and 31 percent of all immigrant women to Canada have a university degree, compared with 18 percent of Canadian-born men and 20 percent of Canadian-born women. Within the class of skilled workers, about 72 percent of immigrants have at least one post-secondary degree. Naomi Alboim, *Fulfilling the Promise: Integrating Immigrant Skills into the Canadian Economy* (Ottawa: Caledon Institute of Social Policy, 2002), 10.

19. "Between $4 and $6 billion is lost to the Canadian economy each year as a result of unrecognized qualifications, and that immigrants are among those who experience the most serious problems in achieving recognition of their learning." Alboim, *Fulfilling the Promise,* 1. See also A. Brouwer, *Immigrants Need Not Apply* (Ottawa: Caledon Institute of Social Policy, 1999).

20. Alboim, *Fulfilling the Promise,* 10. Ekuwa Smith and Andrew Jackson add, "The large gaps in earnings between recent visible minority immigrants and other Canadians *cannot be explained by inferior levels of formal education*" (emphasis added). See Ekuwa Smith and Andrew Jackson, *Does a Rising Tide Lift All Boats? The Labour Market Experiences and Incomes of Recent Immigrants* (Ottawa: Canadian Council on Social Development, 2002), 3. See also Heather Dryburgh, "Immigrant Science and Technology Workers: Employing the Brain Gain?" paper presented at the Congress of Social Sciences and Humanities, Halifax, NS, 2003.

21. Derwyn Sangster, *Assessing and Recognizing Foreign Credentials in Canada: Employers' Views* (Ottawa: Citizenship and Immigration Canada, HRDC, Canadian Chamber of Commerce, and Canadian Labour and Business Centre, 2001), 13. IT employers even expressed dismay at overvaluation of "paper credentials," indicating that relevant experience was much more significant.

22. And yet, for example, Oncu points out that the state participates in deliberately denying skills training to certain people, who are then forced to take low-waged jobs. She points out that "the eligibility requirements of the Employment and Immigration Canada programme exclude many immigrant women," and that these in conjunction with other policies that are "coupled with family responsibilities and obligations severely limit the job opportunities of immigrant women." Oncu, "Informal Economy Participation of Immigrant Women in Canada," 28–29.

23. Additionally, non-quantifiable individual factors such as sex-role socialization and perceptions of social expectations are not included in the human capital analysis. Thea Sinclair, "Women, Work, and Skill," in Nanneke Redclift and Thea Sinclair, eds., *Working Women: International Perspectives on Labour and Gender Ideology* (New York: Routledge, 1991), 6.

24. Sylvia Beyer, Kristina Rynes, Julie Perrault, Kelly Hay and Susan Haller,

"Gender Differences in Computer Science Students," *SIGCSE Bulletin* (2003). The work of Jane Margolis and Allan Fisher also supports these findings.

25. This trend has been well documented in studies of computer use by children. Sherry Turkle's work *Epistemological Pluralism: Styles and Voices Within the Computer Culture* (Cambridge, MA: MIT Epistemology and Learning, 1990) was one of the first to observe differences in computer use by gender, but there is a sizable body of research which has replicated these findings.

26. Sinclair, "Women, Work, and Skill," 5. Sinclair suggests that women's choice to obtain lower skill levels (assuming, for the moment, that women do possess lower skill levels) is a pragmatic one given their anticipated lower earnings and child-care responsibilities.

27. A term used by Ann Crittenden in *The Price of Motherhood: Why the Most Important Job is Still the Least Valued* (New York: Henry Holt and Co., 2001).

28. Sinclair, "Women, Work, and Skill," 5.

29. Ibid., 12.

30. Ibid.

31. Data from Statistics Canada, *CANSIM,* Cross-classified tables 00580701, 00580702.

32. Statistics Canada, "University Qualifications Granted by Field of Study and Sex," cross-classified table 00580602, and "Community College Diplomas in Career Programs by Field of Study and Sex," *Education in Canada* (Ottawa: Statistics Canada, 2001).

33. Picot and Heisz, *The Performance of the 1990s Canadian Labour Market,* 19.

34. Ibid., 23. See also Ross Finnie and Ted Wannell, "The Gender Earnings Gap Amongst Canadian Bachelor's Level University Graduates: A Cross-Cohort Longitudinal Analysis," in Richard Chaykowski and Lisa Powell, eds., *Women and Work* (Kingston: McGill-Queen's University Press, 1999).

35. Tahany Gadalla shows, for example, that women's enrollment in computer science has been dropping consistently since the 1980s. Tahany Gadalla, "Are More Women Studying Computer Science?" *Resources for Feminist Research/ Documentation sur la recherche féministe* 27, no. 1-2 (Spring/Summer 1999), 137–142. Women are least represented in engineering and applied sciences, comprising 21.1 percent of all graduates, and most represented in humanities, education and health professions with 63.5 percent, 71 percent and 72.2 percent of degrees respectively. Statistics Canada, "University Qualifications Granted by Field of Study and Sex" cross-classified table 00580602.

36. Jane Gaskell, Arlene McLaren and Myra Novogrodsky in *Claiming an Education: Feminism and Canadian Schools* (Toronto: Our Schools/Our Selves Education Foundation, 1989) challenge notions of equal opportunity and access in their examination of the Canadian school system. They argue

that systemic and structural barriers, rather than personal deficit, combine to disadvantage females and affect their choices throughout their time in educational institutions. See also Linda Briskin, *Feminist Pedagogy: Teaching and Learning Liberation* (Ottawa: CRIAW/ICREF, 1994). Boel Berner and Ulf Mellström show that engineering schools and workplaces are designed to (re)produce particular kinds of hegemonic masculine values through institutional practice, and actively inhibit women's inclusion. Boel Berner and Ulf Mellström, "Looking for Mister Engineer: Understanding Masculinity and Technology at two Fin de Siécles," in Boel Berner, ed., *Gendered Practices: Feminist Studies of Technology and Society* (Linkoping, Sweden, Department of Technology and Social Change, 1997), 39–68. A more recent study looks at women's experiences in computing at Carnegie Mellon University and identifies several social-structural reasons for women's lack of enrollment and attrition in computer science. Margolis and Fisher, *Unlocking the Clubhouse.*

37. Minna Salminen-Karlsson, "Reforming a Masculine Bastion: State-Supported Reform of Engineering Education," in Berner, ed., *Gendered Practices,* 187–202.

38. Jennifer Stephen, *Access Diminished: A Report on Women's Training and Employment Services in Ontario* (Toronto: Advocates for Community Based Training and Education for Women, 2000), 13.

39. It is beyond the scope of this book to explain in detail the numerous factors that are responsible for women's experiences in technical education. However, in the Bibliography I have suggested sources for further reading.

40. Picot and Heisz, *The Performance of the 1990s Canadian Labour Market,* 12. In a later section the authors note that this is the case in Canada but not in the U.S., where "the wage premium among the more highly educated has risen" (20).

41. The authors argue that this kind of study "on recent graduates as they enter the labour market with relatively little previous experience and freshly minted human capital permits us to isolate the emerging trends which are being driven at the entry-point of the margins of the market." Finnie and Wannell, "The Gender Earnings Gap," 2.

42. Ibid., 38.

43. For example, *Working Woman* magazine, 16 May 2000, cites a study done by Abbot, Langer & Associates for the National Society of Professional Engineers of 1999 median salaries for engineers. Women started at a lower salary than men (average of $3,500 less per year for bachelor's degree), and the pay disparity increased as the level of education increased (average of about $17,000 less for a doctorate). The Society for Technical Communication reports that in 2001, Canadian women working in the field of technical communication (which includes occupations such as technical writers, Web designers and information architects) made approximately $4,000 less per year than men. Interestingly, while salaries in general increased, the disparity between women and men also

increased. In 1997, the STC reported that Canadian women made only $340 less per year on average than men. The U.S. publication *Information Week*'s 2002 salary survey of over 9,000 IT professionals pointed out that not only do women make less than men, the salary gap is growing. They calculate the difference at an average of $7,000 per year.

44. Caroline Weber and Işik Urla Zeytinoğlu, "The Effect of Computer Skills and Training on Salaries," in Chaykowski and Powell, eds., *Women and Work,* 144.

45. K. McMullen, *Skill and Employment Effects of Computer-Based Technology: The Results of Working with Technology* (Ottawa: Canadian Policy Research Networks, 1996), cited in Weber and Zeytinoğlu, "The Effect of Computer Skills and Training on Salaries," in Chaykowski and Powell, eds., *Women and Work,* 143–172.

46. Gordon Betcherman, Darren Lauzon and Norm Leckie, "Technological and Organizational Change and Skill Requirements," Chaykowski and Powell, eds., *Women and Work,* 110.

47. Weber and Zeytinoğlu, "The Effect of Computer Skills and Training on Salaries," 167.

48. Ibid., 169.

49. Eileen Appelbaum's research supports the contentions of Weber and Zeytinoğlu. She writes: "Numerous studies have shown that race, gender, and amount of formal education are important determinants of who receives such 'training' in the U.S." (81). While women were more likely than men to upgrade their skills by funding their own training in community colleges and the like, they did not receive nearly as much on-the-job and employer-funded training as white males. However, Appelbaum stresses, "training outside the workplace is by no means as effective as work-based training in raising either productivity or wages." Eileen Appelbaum, "New Technology and Work Organisation: The Role of Gender Relations," in Belinda Probert and Bruce Wilson, eds., *Pink Collar Blues: Work, Gender and Technology* (Melbourne: Melbourne University Press, 1993), 82. It is not clear here whether this is so because this kind of training is more likely to be acquired by certain groups of people and is thus devalued, or whether employer-funded training does indeed represent superior skill transfer since, as we see in a later section, employer-sponsored training does not always translate to higher wages, status, or skill level.

50. Weber and Zeytinoğlu, "The Effect of Computer Skills and Training on Salaries," 168. Heather Menzies concurs: "Rule-of-thumb self-management and experience-based tacit knowledge," commonly associated with women, "were consistently displaced by credentialled expertise and expert-level software," more commonly associated with men. Heather Menzies, *Whose Brave New World? The Information Highway and the New Economy* (Toronto: Between the Lines, 1996). For a more complete discussion of which kinds of knowledge are associated with male and female ways of knowing, see Lorraine

Code, *What Can She Know?* (Ithaca: Cornell University Press, 1991).

51. Organization for Economic Cooperation and Development, *Technology, Productivity, and Job Creation* (Paris: OECD, 1996).

52. This was characteristic of scholarship published in the 1970s, following Harry Braverman's *Labor and Monopoly Capital: The Degradation of Work in the Twentieth Century* (New York: Monthly Review Press, 1974). For example, A. Zimbalist, "Technology and the Labor Process in the Printing Industry," in A. Zimbalist, ed., *Case Studies on the Labor Process* (London: Monthly Review Press, 1979), 103–126; David Noble, "Social Choice in Machine Design: The Case of Automatically Controlled Machine Tools," also in *Case Studies on the Labor Process,* 18–50. For recent work, see especially Heather Menzies, *Whose Brave New World?* Stanley Aronowitz and William DiFazio note, "All of the contradictory tendencies involved in the restructuring of global capital and computer-mediated work seem to lead to the same conclusion for workers of all collars — that is, unemployment, underemployment, decreasingly skilled work, and relatively lower wages ... High technology will destroy more jobs than it causes ... Technological change and competition in the world market guarantee that increasing numbers of workers will be displaced and that there workers will tend to be rehired in jobs that do not pay comparable wages and salaries." Stanley Aronowitz and William DiFazio, *The Jobless Future: Sci-Tech and the Dogma of Work* (Minneapolis: University of Minnesota Press, 1994), 3–4.

53. Betcherman, Lauzon and Leckie, "Technological and Organizational Change and Skill Requirements."

54. Clerical work is a particularly interesting example of these shaping social relations, for two reasons: its origins as a male-dominated profession, and its location as a site for technological presence (much of the software, such as word processing and spreadsheets, that we use on home computers today began as office technology used by clerical workers). Catherine Cassell takes up these points in her article "A Woman's Place Is at the Word Processor: Technology and Change in the Office," in Jenny Firth-Cozens and Michael West, eds., *Women at Work: Psychological and Organizational Perspectives* (Philadelphia: Open University Press, 1991), 172–184.

55. As Catherine Cassell points out, "The social relations on which secretarial work is based are seen as crucial in defending women's jobs from the negative consequences of advancing technology." Ibid., 179.

56. Aronowitz and DiFazio point out that as a result of "the scientific-technological revolution of our time, which is not confined to electronic processes but [which] has fundamentally altered he forms of work, skill, and occupation[,]... knowledge rather than traditional skill is the productive force." Aronowitz and DiFazio, *The Jobless Future,* 15.

57. Franklin, *The Real World of Technology.*

58. Cassell, "A Woman's Place Is at the Word Processor," 179.

59. Lopez-Bassols, *ICT Skills and Employment,* notes that at present, approximately 60 percent of computer programmers in the U.S. have a bachelor's degree or higher (15). In Canada, information technology occupations are considered to be in the "university" skill requirement group, which means that Statistics Canada assumes that the majority of people in those occupations will be university-educated. He further notes, "Post-secondary education is the main supplier of workers entering IT jobs; nevertheless graduates in computer science and engineering are only a part of all individuals eventually pursuing IT careers." Even more striking, in the UK, "two-thirds of IT workers do not have IT degrees, and two-thirds of IT graduates do not work in IT jobs" (16).

60. There has been a marked increase in the number of commercial technical certifications granted by corporations, business associations, and commercial IT groups. While it remains difficult to empirically gauge the importance of these certifications, in the U.S. in 2000, 1 in 7 positions advertised asked for commercial certifications. Lopez-Bassols, *ICT Skills and Employment,* 17.

61. Gadalla, "Are More Women Studying Computer Science?"

62. For the purposes of this book, I am distinguishing between "computer science," which is regarded as a scientific field of study in which one eventually receives a BSc, MSc or PhD, and "IT," which, though a growing program in many universities and colleges, is not necessarily as closely associated with science. IT courses tend to have an orientation towards finding employment within private corporations and designing, implementing and maintaining particular types of systems such as networks and databases. Computer science tends to have an orientation that is more theoretical, interested in the methods of mathematical computation and places more emphasis on applications such as programming and statistics.

63. See the work of Tove Håpnes and Knut Sørensen, "Competition and Collaboration in Male Shaping of Computing: A Study of Norwegian Hacker Culture," in Keith Grint and Rosalind Gill, eds., *The Gender-Technology Relation: Contemporary Theory and Research* (London: Taylor and Francis, 1995).

64. Eric Dorn Brose, *Technology and Science in the Industrializing Nations, 1500-1914* (New Jersey: Humanities Press, 1998).

65. Sadie Plant, *Zeroes and Ones: Digital Women and the New Technoculture* (New York: Doubleday, 1977), 124-125.

66. Franklin, *The Real World of Technology.*

67. While Franklin's critique is germane, it is worth noting that these kinds of concerns have been common since the Industrial Revolution. Karl Marx, for example, in the late 1850s, predicted systems of production founded on the use of machinery, "consisting of numerous mechanical and intellectual organs, so that the workers themselves are cast merely as its conscious linkages." Karl

Marx, *Grundrisse (Foundations of the Critique of Political Economy)* (1857–58; reprint, Harmondsworth, UK: Penguin, 1973), 692.

68. Robert Laubacher and Thomas Malone, "Retreat of the Firm and Rise of Guilds: The Employment Relationship in an Age of Virtual Business." *MIT Initiative on Inventing the Organizations of the Twenty-first Century Working Paper* (Cambridge, MA: MIT, 2002).

Chapter 3: Great Promises versus Material Realities

1. Y2K refers to the anticipated problem of computers not being able to correctly identify the year 2000. Because computer clocks had been set with only a two-digit date (for example, 1993 would be "93"), people worried as 2000 approached that there would be a technological meltdown with far-reaching consequences. For example, a computer that had a date that it didn't understand might shut down or not work properly, and if this computer were an important one, such as a bank's data server, then it could cause significant problems. Large numbers of IT workers were hired on a contract basis to update computer systems to prevent this from occurring. Many spent their New Year's Eve in 1999 standing by in case of emergency. Whether because of the hard work of these people, or because computers managed to understand 2000 just fine, Y2K was one of the biggest anticlimaxes in IT history.

2. Human Resources Development Canada, *The Call Centre Sector in Canada* (Ottawa: HRDC, 1999), 6.

3. The seven assertions of information revolutionaries are paraphrased from Nick Dyer-Witheford, *Cyber-Marx: Cycles and Circuits of Struggle in High-Technology Capitalism* (Urbana: University of Illinois Press, 1999), 22–26.

4. Ibid., 25.

5. Ibid., 97.

6. Melanie Stewart Millar, *Cracking the Gender Code: Who Rules the Wired World?* (Toronto: Second Story Press, 1998; now available from Sumach Press, Toronto).

7. D. Whitney, Jr., *I'm Glad I'm a Boy! I'm Glad I'm a Girl!* (New York: Simon and Schuster, 1970), quoted in Jane Margolis and Allan Fisher, *Unlocking the Clubhouse: Women in Computing* (Cambridge, MA: MIT Press, 2002), 2.

8. Gordon Betcherman and Kathryn McMullen, *Impact of Information and Communication Technologies on Work and Employment in Canada, Dicussion Paper* (Ottawa: Canadian Policy Research Networks, 1998), 6.

9. Celia Stanworth, "Telework and the Information Age," *New Technology, Work, and Employment* 13, no. 1 (1997), 54.

10. Dyer-Witheford, *Cyber-Marx,* 144.

11. "In 2001 ... the ICT services industries kept growing, up by 12.3%. As a result, over the 1997–2001 period, the ICT services industries have grown faster than the ICT manufacturing and wholesaling industries (15.6% per year versus 10.6% and 10.4% respectively). *Information and Communications Technologies Industry, April 2002 Industry Innovation Profile, Canada's Innovation Strategy* (Ottawa: Ministry of Industry). Avaliable on-line from www.innovationstrategy.gc.ca.

12. Carla Freeman, *High Tech and High Heel in the Global Economy: Women, Work and Pink-Collar Identities in the Caribbean* (Durham: Duke University Press, 2000).

13. See Stanworth, "Telework and the Information Age," 51–62.

14. Laura Miller, "Women and Children First: Gender and the Settling of the Electronic Frontier," in James Brook and Iain Boal, eds., *Resisting the Virtual Life: The Culture and Politics of Information* (San Francisco: City Lights Books, 1995).

15. Don Little, "Employment and Remuneration in the Services Industries Since 1984," *Statistics Canada Analytical Paper Series No. 24* (Ottawa: Statistics Canada and Industry Canada, 1999).

16. Janice McLaughlin, "Gendering Occupational Identities and IT in the Retail Sector," *New Technology, Work, and Employment* 14, no. 2 (1999), 143–156.

17. However, it is worth noting that types of service work differ in their traditional demographic makeup. If a service worker is white, middle class and male he is very likely to be young. Service work that depends on visibility, such as serving food, is more likely to be performed by younger white workers. Service work that depends on invisibility, such as cleaning hotel rooms or working in a kitchen, is more likely to be performed by workers of colour and recent immigrants. See, for example, Amel Adib and Yvonne Guerrier, "The Interlocking of Gender with Nationality, Race, Ethnicity and Class: The Narratives of Women in Hotel Work," *Gender, Work and Organisation* 10, no. 4 (August 2003), 413–432, for a treatment of the intersecting relations that shape experiences of service work.

18. HRDC, *The Call Centre Sector in Canada,* 1.

19. Phil Taylor and Peter Bain, "'An Assembly Line in the Head': Work and Employee Relations in the Call Centre," *Industrial Relations Journal* 30, no. 2 (1999), 101–117.

20. Jane Jenson, "The Talents of Women, the Skills of Men: Flexible Specialization and Women," in S. Wood, ed., *The Transformation of Work? Skill, Flexibility, and the Labour Process* (London: Unwin Hyman, 1989), 155; McLaughlin, "Gendering Occupational Identities and IT in the Retail Sector."

21. Jamie Carson, *An Updated Look at the Computer Services Industry* (Ottawa: Statistics Canada, Service Industries Division, 2001), 8; HRDC, "The Call Centre Sector in Canada."

22. Information and Communications Technologies Industry, April 2002 *Industry Innovation Profile, Canada's Innovation Strategy.* As Jamie Carson notes, "With an output growth rate of six times that for the overall economy since 1992, to say that the computer services industry is growing in Canada would be an understatement." Carson, *An Updated Look at the Computer Services Industry.*

23. Robert Laubacher and Thomas Malone, "Retreat of the Firm and the Rise of Guilds: The Employment Relationship in an Age of Virtual Business." *MIT Initiative on Inventing the Organizations of the Twenty-first Century Working Paper* (Cambridge, MA: MIT, 2002), 4.

24. Ibid.

25. "While employees may be less secure, their work is often more interesting and fulfilling ... Workers now bear the brunt of their company's or the economy's misfortunes in a way they did not under the old system." Ibid., 13–14.

26. Information and Communications Technologies Industry, April 2002 *Industry Innovation Profile, Canada's Innovation Strategy.*

27. The 2001 data are from Statistics Canada *Survey of Labour and Income Dynamics* (SLID).

28. Betcherman and McMullen, *Impact of Information and Communication Technologies on Work and Employment in Canada,* 12. Their data from 1992–1994 show that professional jobs represented approximately 57 percent of the new IT jobs created, while intermediate jobs (a notch above unskilled) were lost at almost the same rate, 60 percent.

29. Taylor and Bain, "'Assembly Line in the Head,'" 102.

30. Ibid., 106.

31. "Canadian Colleges and Institutes Play Critical Role in Preparing the Workforce for Competitive Internet Economy," press release from the Association of Canadian Community Colleges, 23 August 2001.

32. Richard Fung, quoted in Ryan B. Patrick, "Vets Help Grads Navigate Market," *IT World Canada Skills Management Newsletter,* 25 July 2003.

33. The comptometer was patented in 1887 and remained in use until the 1970s. It was a very popular and commonplace piece of machinery in offices during this time. Though operation of this machine was complex and required extensive training, an educated operator could manage columns of numbers with great speed. Sharon Hartman Strom, "'Machines Instead of Clerks': Technology and the Feminization of Bookkeeping, 1910-1950," in Heidi Hartmann, ed., *Computer Chips and Paper Clips: Technology and Women's Employment.* Vol. 2: *Case Studies and Policy Perspectives* (Washingon, DC: National Academy Press, 1987), 63–97.

Chapter 4: New Work versus Same Old, Same Old

1. Toronto Organizing for Fair Employment (TOFFE) is leading the protest against Rogers' work practices. See their Web site at www.toffeonline.org. Also see the Ontario Federation of Labour's report, *The People's Charter: Employment Standards,* which notes, "Workers at the bottom of [the] subcontracting chain are more vulnerable to employment standards violations and receive much lower wages and few benefits. A good example of this is the case of workers who sold Rogers Cable Internet and Cable TV subscriptions door-to-door in the Toronto area. Although they were trained by Rogers' staff in a Rogers' facility, and carried Rogers' ID badges, they were hired by an 'independent' subcontractor. After working for several weeks, these workers were paid only a small fraction of what they had earned, or nothing at all. The subcontractor has gone out of business, and Rogers is denying any responsibility for these lost wages." The report is available on-line from www.ofl–fto.on.ca/library/5-Employment-Standards.pdf.
2. For example, as a result of a suit launched in Australia by a broadcasting company, a precedent-setting ruling by the Australian Industrial Relations Commission now allows companies to prohibit employee access to union e-mail. Julian Bajkowski, *Computerworld Today* (Australia, on-line edition), 28 July 2003.
3. Nick Dyer-Witheford, *Cyber-Marx: Cycles and Circuits of Struggle in High-Technology Capitalism* (Urbana: University of Illinois Press, 1999), 9.
4. "Post-Fordism" refers to a system of production that followed from "Fordism." Fordism, named after the production practices of Henry Ford, emphasized mass production combined with mass consumption. Post-Fordism emphasizes smaller, often dispersed or fragmented units of production and so-called flexible specialization. A typical Fordist site of production is a large unionized factory that is governed by state labour regulations. A typical post-Fordist site of production is a small subcontracted shop in a low-cost, poorly regulated geographic region.
5. As Leah Vosko indicates, the standard employment relationship was a normative paradigm of employment that necessitated the social context of the "postwar welfare state where workers successfully secured associational rights, where collective bargaining rights were becoming the norm, and where rates of unionization were relatively high." Leah Vosko, *Temporary Work: The Gendered Rise of a Precarious Employment Relationship* (Toronto: University of Toronto Press, 2000), 25.
6. "People then find themselves working directly for computer systems, as extensions of their operating software, with no opportunity for advancement or involvement past the silicon curtain descending inside the operating system."

Heather Menzies, *Whose Brave New World? The Information Highway and the New Economy* (Toronto: Between the Lines, 1996), 36. This view of human as extension of machine is in direct contrast to Sandy Stone's positive vision of the advantages of technological prosthesis, technology as an extension of humans, in *The War of Desire and Technology at the Close of the Mechanical Age* (Cambridge, MA: MIT Press, 1995).

7. Menzies, *Whose Brave New World?* 61.

8. Ibid.

9. Sheila Allen and Carol Wolkowitz describe the consequences of this: "Ironically enough, ["modernization"] frequently involves a return to methods of production previously discussed as archaic. Whereas firms once sought to regularise the employment of their work-force as a way of increasing worker commitment and productivity, those directly employed by large firms and state corporations and services are increasingly being asked to accept employment on renegotiated terms." Sheila Allen and Carol Wolkowitz, *Homeworking: Myths and Realities* (London: MacMillan Education, 1987), 165.

10. As, for example, Vosko also suggests.

11. Anna Pollert, "Dismantling Flexibility," *Capital and Class* 34 (Spring 1988), 43. Also see, Ursula Franklin, *The Real World of Technology* (Toronto: House of Anansi Press, 1999), and Melanie Stewart Millar, *Cracking the Gender Code: Who Rules the Wired World?* (Toronto: Second Story Press, 1998; now available from Sumach Press, Toronto).

12. Pollert, "Dismantling Flexibility," 50.

13. Telework is defined as "a work arrangement in which organizational employees regularly work at home, or at a remote site, one or more complete workdays a week" by Linda Duxbury, Christopher Higgins and Derrick Neufeld, "Telework and the Balance Between Work and Family: Is Telework Part of the Problem or Part of the Solution?" in Magid Igbaria and Margaret Tan, eds., *The Virtual Workplace* (Hershey, PA: Idea Group Publishing, 1998), 221.

14. The experiences of individual teleworkers are influenced by their personal trajectory into the workforce. That experience and its evaluation range from the life-enhancing, by virtue of the freedoms offered, to the exploitative, when there is no other option. In view of the complex and disparate nature of teleworking, both economically and sociologically, it is important to be cautious in making both judgements and predictions about its significance. Leslie Haddon and Roger Silverstone, "Home Based Telework," in William Dutton, ed., *Society on the Line: Information Politics in the Digital Age* (New York: Oxford University Press, 1999), 164. Barbara Orser and Mary Foster concur: "The reasons for choosing to work in a home-based business are as complex and diverse as the people involved in this kind of enterprise" in *Home Enterprise: Canadians and Home-Based Work* (Abbotsford, BC: Home-Based Business Project Committee, 1992), 92.

15. Menzies, *Whose Brave New World?* 77.

16. Haddon and Silverstone, "Home Based Telework," 163.

17. Celia Stanworth, "Telework and the Information Age," *New Technology, Work, and Employment* 13, no. 1 (1997), 57. Also Diane Bailey and Nancy Kurland, "A Review of Telework Research: Findings, New Directions, and Lessons for the Study of Modern Work," *Journal of Organizational Behavior* 23 (2002), 383–400.

18. Stanworth, "Telework and the Information Age," 57.

19. See Eileen Appelbaum, "New Technology and Work Organisation: The Role of Gender Relations," in Belinda Probert and Bruce Wilson, eds., *Pink Collar Blues: Work, Gender and Technology* (Melbourne: Melbourne University Press, 1993), and Sally Hacker, "Sex Stratification, Technology, and Organizational Change: A Longitudinal Case Study AT&T," in Dorothy Smith and Susan Turner, eds., *Doing It the Hard Way: Investigations of Gender and Technology* (London: Unwin Hyman, 1990), 45–67.

20. Bailey and Kurland, "A Review of Telework Research," 386. Also Diane-Gabrielle Tremblay, "Balancing Work and Family with Telework: Organizational Issues and Challenges for Women and Managers," *Women in Management Review* 17, nos. 3/4 (157), 157–170.

21. William H. Dutton, ed., *Society on the Line: Information Politics in the Digital Age* (New York: Oxford University Press, 1999), 117–118.

22. David Grimshaw and Sandy Kwok, "The Business Benefits of the Virtual Organization," in Igbaria and Tan, eds., *The Virtual Workplace,* 64–66.

23. In terms of the advantages of telework in particular, William Dutton argues that ICTs (information and communications technologies) "can overcome many constraints of time and distance to place organizations in the presence of critical information, people, and services. This can liberate creative, administrative, management, production, and distribution resources and processes from the constraints imposed by traditional organizational structures." Dutton, ed., *Society on the Line,* 114.

24. Tremblay, "Balancing Work and Family with Telework," 161–162.

25. And, as Dutton cautions, "economic payoffs and competitive advantages from ICTs are not inevitable ... technical innovations also require social and organizational innovations." Dutton, ed., *Society on the Line,* 119. For more on home-based work and child care, see Tremblay, "Balancing Work and Family with Telework," 166.

26. Grimshaw and Kwok, "The Business Benefits of the Virtual Organization," 67.

27. "It is something of a puzzle why the opportunities created by information technologies have not had a more dramatic effect on management and work methods. In part, the answer is that previously established institutional structures and old forms of work organisation continue to exert an important

influence as new technologies are introduced." Appelbaum, "New Technology and Work Organisation," 60.

28. Juliet Webster, *Office Automation: The Labour Process and Women's Work in Britain* (New York: Harvester Wheatsheaf, 1990), 126.

29. "With few exceptions ... accounts of increased productivity under telework are derived from self-report data ... In [one] sample, 48 per cent of teleworkers report increased hours when working at home, which raises the possibility that the teleworkers may be conflating improved productivity with an increase in the absolute amount of work performed ... In [Olson's] survey, 67 per cent of the people who work at home report increased productivity; among them, 40 per cent report that they work too much. Self-report data, for a variety of reasons, fails to provide convincing support for productivity claims." Bailey and Kurland, "A Review of Telework Research," 389.

30. Ralph Westfall, "The Microeconomics of Remote Work," in Igbaria and Tan, eds., *The Virtual Workplace.*

31. Ibid., 260.

32. Leslie Willcocks and Stephanie Lester, "Information Technology: Transformer or Sink Hole?" *Beyond the IT Productivity Paradox* (New York: John Wiley and Sons, 1999), 1.

33. Allen and Wolkowitz, *Homeworking: Myths and Realities,* 175.

34. Ibid., 178.

35. Dutton, ed., *Society on the Line,* 131. See also Ronald Richardson and Andrew Gillespie, "The Impact of Remote Work on Employment Location and Work Processes," in Dutton, ed., *Society on the Line,* 165.

36. Sitel, "Leveraging the Offshore Market for Contact Center Solutions: Interview with Fred Shadding, VP, Offshore Solutions and Alliances." Sitel PR material available on-line from www.sitel.com/enu/Interview04. stm.

37. Pollert, "Dismantling Flexibility," 64.

38. Trevor Haywood, "Global Networks and the Myth of Equality," in Brian Loader, ed., *Cyberspace Divide: Equality, Agency, and Policy in the Information Society* (New York: Routledge, 1998).

39. Sitel, "Leveraging the Offshore Market for Contact Center Solutions."

40. Donna Haraway, "A Cyborg Manifesto: Science, Technology, and Socialist-Feminism in the Late Twentieth Century," in *Simians, Cyborgs, and Women: The Reinvention of Nature* (New York: Routledge, 1991), 153.

41. For an interesting discussion of this idea demonstrated through representation, see Lisa Nakamura, "'Where Do You Want to Go Today?' Cybernetic Tourism, the Internet, and Transnationality," in Beth Kolko, Lisa Nakamura and Gilbert B. Rodman, eds., *Race in Cyberspace* (New York: Routledge, 2000), 15–26.

42. Carla Freeman, *High Tech and High Heels in the Global Economy: Women, Work, and Pink-Collar Identities in the Caribbean* (Durham, NC: Duke University Press, 2000).

43. A play on *maquiladoras,* factories which are set up along the U.S.–Mexico border, owned by non-Mexican corporations and which produce goods for U.S. consumption. *Maquiladora* workers are primarily female, low-paid and low-status. The cheapness of the goods they export to the U.S. depends, in part, on the cheapness of this labour.

44. Melanie Stewart Millar also makes this point in *Cracking the Gender Code: Who Rules the Wired World?*

45. "Workers Rights," WashTech. Available on-line from www.washtech.org/wt/rights.

46. "Why We Need Alliance@IBM." Available on-line from www.allianceibm.org/why.html.

CHAPTER 5: LOOKING AHEAD

1. Mindy Gewirtz and Ann Lindsey, *Women in the New Economy: Insights and Realities* (Brookline, MA: GLS Consulting, 2000), 3.

2. Steve Maich, "How Greed Took Down Nortel," *Financial Post* (on-line edition), 29 April 2004.

3. Janet McFarland, Karen Howlett and Dave Ebner, "How Nortel Lost the Race to Redemption," *The Globe and Mail,* 1 May 2004, B5.

4. Adam Grachnik, "Girls Gone Filed, " *ITBusiness.ca* (on-line edition), 7 May 2004.

5. Karen Hughes, Graham S. Lowe and Grant Schellenberg, *Men's and Women's Quality of Work in the New Canadian Economy* (Ottawa: Canadian Policy Research Network, 2003).

6. Statistics Canada, *Centre for Education Statistics* (Ottawa, 2004).

7. Jane Margolis and Allan Fisher, *Unlocking the Clubhouse: Women in Computing* (Cambridge, MA: MIT Press, 2002).

8. *The Labour Force Survey,* from which this data are taken, uses a different and earlier occupational classification (Standard Occupational Classification 91 [SOC91]) than the *Census* (National Occupational Classification for Statistics, 2001 [NOC-S 2001]). As such, professions are not precisely comparable, but with a few exceptions (such as Web designers, which in the LFS are classified as graphic designers) are a close approximation.

9. This is not a complete list of all IT professions, but rather only the ones classified as "Computer and Information Systems Professionals" using the National Occupational Classification for Statistics categorization, C047,

C071-C075. See also note 8. I use the term "visible minority" advisedly; it is not my preferred term. Statistics Canada uses the category to denote any person or group identified as non-Caucasian. Aboriginal people are not considered to be visible minorities by Statistics Canada; they are included in the "non-visible minority" grouping. This chart compares four groups of women: visible minority immigrant women; visible minority non-immigrant (Canadian-born) women; non-visible minority, immigrant women; and non-visible minority, non-immigrant (Canadian-born) women. The charts also include non-visible minority, non-immigrant (Canadian-born) men as a comparison.

10. Hughes, Lowe, Schellenberg, *Men's and Women's Quality of Work in the New Canadian Economy.*

11. Ibid.

12. Daniel H. Pink, "The New Face of the Silicon Age," *Wired,* February 2004, 98.

Glossary

Access: The ability, right or privilege to use technology. This generally requires various components such as technological literacy, economic and social access to the infrastructure and equipment, and time and space in which to use it.

Agency: The ability of a person to make choices from the options that she/he feels are available, rendering her/him an active participant, rather than a passive observer.

Automation: The transfer of task performance from human workers to machines, or the control and organization of work processes by machines.

Call centre: An organization that exists to process telephone communication. Inbound call centres typically provide services such as customer support or information, while outbound call centres typically sell or market products or services. In both cases, information and communications technologies are merged to organize and route calls and to supervise and record employees' work.

Computer networking: The linkage of one computer to another, of a personal computer to a larger system or of larger systems to one another.

Computer science: An academic discipline concerned primarily with the development of the conceptualization, creation, development, implementation and manipulation of computer systems and languages.

Computerization: The conversion of a manual work process to a computer-based process, or the implementation of computers in a workplace.

Computers: Machines designed to facilitate and speed up computation, electronic information storage and logical processes. Also a term used to describe the workers, usually women, who computed ballistics tables and other mathematical equations for the military during the Second World War.

Cyberculture: A culture oriented around the actual use and future possibilities of technology, particularly human-technology interaction. This term can also refer to the culture generated by on-line interaction.

Cyberfeminism: A branch of feminism interested in the role that technology, particularly information technology, can play in facilitating women's social equality, activism and equity. Cyberfeminists are also concerned with issues of women's access to technology globally and the creation of women's communities on-line.

CYBERSPACE: This term primarily refers to the virtual "space" created by the internet, and is often conceptualized as an ethereal medium of digitized information. "Cyberspace" is also often used to refer to any imagined space of electronic technological interaction.

CYBORG: A human-machine hybrid. Cyborgs can be imaginary creatures or they can represent real human-machine interfaces. They symbolize the blurring of boundaries between "nature" and "civilization," or "organic" and "inorganic," as well as the creation of entirely new organisms and items from the combination of elements.

DATA ENTRY: Both the act of inputting data and information as well as the jobs associated with this task. In general, data entry jobs are considered low in skill and status. As such they are often allotted to marginalized groups of workers.

DEFICIT MODEL: The notion that females have some inherent deficit, either biological or socially constructed, which prevents or inhibits their entry into technological or other male-dominated fields. This model locates the responsibility for educational and occupational segregation and stratification with females, rather than the conditions and structures of society.

DESKILLING: The downgrading, devaluing or elimination of particular workplace skills, usually accompanied by the deliberate implementation of specific technologies, attempts to quantify task performance and a change to the job's status.

DIGITAL DIVIDE: An expression of the global division between those with the social and economic privilege of using technology and those without. The digital divide is not just about technological literacy but about access and ownership of technological resources.

ELECTRONIC COMMERCE (ALSO KNOWN AS "E-COMMERCE"): A form of commercial exchange conducted using electronic or virtual technologies, such as making a purchase through the Internet by using a credit card.

ELECTRONIC COTTAGE: A metaphor used to describe the technologically facilitated transfer of labour from the "public" workplace into the "private" home, and the accompanying creation of electronically linked communities.

ELECTRONIC SURVEILLANCE: The observation and supervision of others through means of electronic devices, such as video cameras or software applications that measure keystrokes.

EMPLOYEE MONITORING: Observation of employees by an employer for the purposes of quantifying and regulating their work. This is often done by electronic means, such as listening in on telephone calls, or measuring the time it takes to complete a computer transaction (see Electronic surveillance).

HIGH TECHNOLOGY: Very advanced or complex technological components or processes, usually used in scientific, medical or engineering-based industries.

INDUSTRIAL SEGREGATION: The division of groups of people (i.e., by gender) into different industries. These industries may be unequally valued as well as divided or be characterized by differences in pay, status, unionization rates, regulatory protection and quality of working conditions.

INFORMATION POVERTY: The state of not having access to or the skills to use information technologies or the data they provide.

INFORMATION TECHNOLOGY (IT): Technological objects and processes used for the creation, manipulation, communication and storage of information and data. Also often called information and communications technology (ICTs).

LUDDITES: Workers in nineteenth-century England who destroyed machinery as a form of protest against poor working conditions and unemployment. The term has since come to mean any opponent of technology, though this was not explicitly what the original Luddites were rebelling against.

NETWORKS: Interconnected, usually non-linear, systems, such as computer networks.

NEW ECONOMY: A term used to describe a post-industrial economy that is technologically based. The "new economy" often suggests a new way to generate wealth and resources.

OCCUPATIONAL SEGREGATION: The division of groups of people into different occupations. These occupations may be horizontally divided, such as different occupations at the same level within an organization, or they may be organized hierarchically. They are likely to be characterized by differences in pay, status, unionization rates, regulatory protection and quality of working conditions.

PINK COLLAR: A term that refers to a particular domain of female-dominated service-based employment — clerical work, for example.

REPRODUCTIVE TECHNOLOGY: Technology intended either to facilitate or prevent reproduction. The majority of reproductive technologies are used by women. The application of reproductive technologies, and access to reproductive choice, depend heavily on social and economic privilege.

SHADOW WORK: Work, often enabled by technology, that used to be paid work but has been downloaded on to consumers; for example, self-checkouts, self-serve gas and automated banking. This results in both the elimination of jobs as well as an increasing load of unpaid work.

SPEED-UP: The process of work intensification whereby increased speed of task performance is demanded of an employee by an employer. This may be accompanied by the implementation of particular technologies that monitor tasks or speed work up.

SUBSISTENCE TECHNOLOGY: Technology used as part of subsistence-based agricultural production or rural living, such as water pumps.

TECHNOLOGICAL RESISTANCE: Resistance to technology, or the use of technology to resist something else. For example, the development of open source code is an act of technological resistance against the monopoly of Microsoft operating systems.

TECHNOLOGICAL UNEMPLOYMENT: Unemployment that results directly from the implementation of workplace technologies. This can be an intentional decision on the part of an employer, usually to maximize profits, or can be an unforeseen consequence of such technological developments.

TECHNOSCIENCE: A term that references the connection and mutual reinforcement between technology and science.

TELEWORK: Work that is facilitated by information and communications technologies and that which is performed at a location other than the premises of the employer. Telework is often distinguished from telecommuting by status and compensation; telecommuting refers to higher paid professional work that the worker chooses to perform at home.

UNPAID LABOUR: Work done without pay or an employment relationship. Typically this is thought of as domestic work, such as child care or house cleaning, but can include any labour that is performed for free.

UPSKILLING: A perceived or actual increase in skill level required for a task or occupation.

VIRTUAL REALITY: A simulated three-dimensional space or environment created and experienced by users through various technological interfaces.

Selected Bibliography

Adler, Nancy J., and Dafna N. Izraeli, eds. *Competitive Frontiers: Women Managers in a Global Economy.* Cambridge, MA: Blackwell Publishers, 1994.

Alboim, Naomi. *Fulfilling the Promise: Integrating Immigrant Skills into the Canadian Economy.* Ottawa: Caledon Institute of Social Policy, 2002.

Allen, Sheila, and Carol Wolkowitz. *Homeworking: Myths and Realities.* London: MacMillan Education, 1987.

Amadeo, Edward J., and Susan Horton. *Labour Productivity and Flexibility.* New York: St. Martin's Press, 1997.

Amott, Teresa, and Julie Matthaei. *Race Gender and Work: A Multicultural Economic History of Women in the United States.* Boston: South End Press, 1996.

Arai, A. Bruce. "Self-Employment as a Response to the Double Day for Women and Men in Canada." *Canadian Review of Sociology and Anthropology* 37, no. 2 (May 2000): 125–142.

Armstrong, Pat, and Hugh Armstrong. *The Double Ghetto: Canadian Women and Their Segregated Work.* 3d ed. Toronto: McClelland and Stewart, 2001.

Aronowitz, Stanley, and William DiFazio. *The Jobless Future: Sci-Tech and the Dogma of Work.* Minneapolis: University of Minnesota Press, 1994.

Avon, Emmanuelle. *Human Resources in Science and Technology in the Services Sector.* Ottawa: Statistics Canada, Services, Science and Technology Division, 1996.

Baker, Michael, and Nicole Fortin. *The Gender Composition and Wages: Why Is Canada Different From the United States?* Ottawa: Statistics Canada, Business and Labour Market Analysis Division, and Human Resources Development Canada, Applied Research Branch, 2000.

Benston, Margaret. "The Political Economy of Women's Liberation." *Monthly Review* 21 (September 1969).

Berner, Boel, ed. *Gendered Practices: Feminist Studies of Technology and Society.* Linkoping, Sweden: Department of Technology and Social Change, 1997.

Betcherman, Gordon, and Kathryn McMullen. *Impact of Information and Communication Technologies on Work and Employment in Canada. Discussion Paper.* Ottawa: Canadian Policy Research Networks, 1998.

Bibby, Andrew. "Teleworking and the Trade Union Movement." Available on-line at http://www.eclipse.co.uk/pens/bibby/workers.html. Originally published in *The Journal of World Transport Police and Practice* 12, no. 1/2 (1996).

Brand, Dionne. "'We Weren't Allowed to Go Into Factory Work Until Hitler Started the War': The 1920s to the 1940s." In Peggy Bristow, ed., *We're Rooted Here and They Can't Pull Us Up: Essays in African Canadian Women's History.* Toronto: University of Toronto Press, 1994.

Braverman, Harry. *Labour and Monopoly Capital: The Degradation of Work in the Twentieth Century*. New York: Monthly Review Press, 1974.

Brouwer, A. *Immigrants Need Not Apply.* Ottawa: Caledon Institute of Social Policy, 1999.

Burnell, Barbara S. *Technological Change and Women's Work Experience: Alternative Methodological Perspectives*. Westport, CT: Bergin and Garvey, 1993.

Canadian Advisory Council on Science and Technology. *Stepping Up: Skills and Opportunities in the Knowledge Economy.* Ottawa: Industry Canada, 2000.

Carson, Jamie. *An Updated Look at the Computer Services Industry.* Ottawa: Statistics Canada, Service Industries Division, 2001.

Cassell, Catherine. "A Woman's Place is at the Word Processor: Technology and Change in the Office." In Jenny Firth-Cozens and Michael West, eds., *Women at Work: Psychological and Organizational Perspectives*. 172–184. Philadelphia: Open University Press, 1991.

Cassell, Justine, and Henry Jenkins, eds. *From Barbie To Mortal Kombat: Gender and Computer Games*. Cambridge, MA: MIT Press, 1998.

Catalyst. *Closing the Gap: Women's Advancement in Corporate and Professional Canada*. Ottawa: Catalyst and The Conference Board of Canada, 1997.

Chaykowski, Richard, and Lisa Powell, eds. *Women and Work*. Kingston: McGill-Queen's University Press, 1999.

Cockburn, Cynthia. *Machinery of Dominance: Women, Men and Technical Knowhow.* London: Pluto Press, 1985.

Cockburn, Cynthia, and Susan Ormrod. *Gender and Technology in the Making.* London: Sage Publications, 1993.

Crittenden, Ann. T*he Price of Motherhood: Why the Most Important Job Is Still the Least Valued.* New York: Henry Holt and Co., 2001.

Date-Bah, Eugenia, ed. *Promoting Gender Equality at Work: Turning Vision into Reality in the Twenty-First Century.* New York: Zed Books, 1997.

Dickinson, Paul, and George Sciadas. *Access to the Information Highway*. Ottawa: Statistics Canada, Services, Science, and Technology Division, 1996.

—. *Access to the Information Highway: The Sequel.* Ottawa: Statistics Canada, Services, Science and Technology Division, 1997.

Dickinson, Paul, and Jonathan Ellison. *Getting Connected or Staying Unplugged: The Growing Use of Computer Communications Services*. Ottawa: Statistics Canada, Service Industries Division, 1999.

Disco, Cornelis, and Barend van der Meulen, eds. *Getting New Technologies Together: Studies in Making Sociotechnical Order.* New York: Walter de Gruyter, 1998.

Drolet, Marie. *Wives, Mothers, and Wages: Does Timing Matter?* Ottawa: Statistics Canada, Business and Labour Market Analysis Division, 2002.

Drolet, Marie. *The Persistent Gap: New Evidence on the Canadian Gender Wage Gap.* Ottawa: Statistics Canada, Income Statistics Division, 2001.

Dryburgh, Heather. *Changing Our Ways: Why and How Canadians Use the Internet.* Ottawa: Statistics Canada, 2000.

Duffy, Ann, and Norene Pupo. *Part-Time Paradox: Connecting Gender, Work, and Family.* Toronto: McClelland and Stewart, 1992.

Dutton, William H., ed. *Society on the Line: Information Politics In the Digital Age.* New York: Oxford University Press, 1999.

Dyer-Witheford, Nick. *Cyber-Marx: Cycles and Circuits of Struggle in High-Technology Capitalism.* Chicago: University of Illinois Press, 1999.

Eisenstein, Zillah. *Global Obscenities: Patriarchy, Capitalism and the Lure of Cyberfantasy.* New York: New York University Press, 1998.

Ertl, Heidi. *Beyond the Information Highway: Networked Canada.* Ottawa: Statistics Canada, Science, Innovation and Electronic Information Division, and Minister of Industry, 2001.

Fawcett, Gail. *Bringing Down the Barriers: The Labour Market and Women with Disabilities in Ontario.* Ottawa: Canadian Council on Social Development, 2000.

Feenburg, Andrew. *Questioning Technology.* London: Routledge, 1999.

Firth-Cozens, Jenny, and Michael West, eds. *Women at Work: Psychological and Organizational Perspectives.* Philadelphia: Open University Press, 1991.

Floyd, Michael, ed. *Information Technology Training for People with Disabilities.* London, UK: Jessica Kingsley Publishers, 1993.

Franklin, Ursula. *The Real World of Technology.* Toronto: House of Anansi Press, 1999.

Friedan, Betty. *The Feminine Mystique.* New York: Dell, 1963.

Fuller, Margaret. *Woman in the Nineteenth Century.* 1855. Reprint, New York: W.W. Norton, 1971.

Gadalla, Tahany. "Are More Women Studying Computer Science?" *RFR/DRF* 27 no. 1/2 (Spring/Summer 1999): 137–142.

Galarneau, Diane, Howard Krebs, René Morissette and Xuelin Zhang. *The Quest for Workers: A New Portrait of Job Vacancies in Canada.* Ottawa: Statistics Canada and Human Resources Development Canada, 2001.

Gaskell, Jane, Arlene McLaren, and Myra Novogrodsky. *Claiming an Education: Feminism and Canadian Schools.* Toronto: Our Schools/Our Selves Education Foundation, 1989.

Greer, Germaine. *The Female Eunuch.* London: Paladin, 1970.

Grint, Keith, and Rosalind Gill, eds. *The Gender-Technology Relation: Contemporary Theory and Research.* London: Taylor and Francis, 1995.

Grint, Keith, and Steve Woolgar. *The Machine At Work: Technology, Work and Organization.* Cambridge, UK: Polity Press, 1997.

Gurstein, Penny. *Wired to the World, Chained to the Home: Telework in Daily Life.* Vancouver: University of British Columbia Press, 2001.

Hacker, Sally. *Pleasure, Power, and Technology: Some Tales of Gender, Engineering, and the Cooperative Workplace.* Boston: Unwin Hyman, 1989.

—. "Sex Stratification, Technology, and Organizational Change: A Longitudinal Case Study AT&T." In Dorothy Smith and Susan Turner, eds., *Doing It the Hard Way: Investigations of Gender and Technology.* 45–67. London: Unwin Hyman, 1990.

Handler, Joel, and Lucie White. *Hard Labor: Women and Work in the Post-Welfare Era.* New York: M.E. Sharpe, 1999.

Haraway, Donna. *Simians, Cyborgs, and Women: The Reinvention of Nature.* New York: Routledge, 1991.

—. *Modest_Witness@Second_Millenium.FemaleMan©_Meets_OncoMouse*^{tm}*: Feminism and Technoscience.* New York: Routledge, 1996.

Harding,Sandra. *Is Science Multicultural? Postcolonialisms, Feminisms, and Epistemologies.* Bloomington, IN: Indiana University Press, 1998.

Hartsock, Nancy. *The Feminist Standpoint Revisited and Other Essays.* Boulder, CO: Westview Press, 1998.

Haywood, Trevor. *Info-Rich, Info-Poor: Access and Exchange in the Global Information Society.* London: Bowker-Saur, 1995.

Herman, Bohuslav, and Wim Stoffers, eds. *Unveiling the Informal Sector: More than Counting Heads.* Aldershot, UK: Avebury, 1996.

Holmes, David, ed. *Virtual Politics: Identity and Community In Cyberspace.* London: Sage Publications, 1997.

Horowitz, Roger, and Arwen Mohun, eds. *His and Hers: Gender, Consumption and Technology.* Charlottesville: University Press of Virginia, 1998.

Human Resources Development Canada and Industry Canada. *Achieving Excellence: Investing in People, Knowledge, and Opportunity.* Ottawa: Industry Canada, 2001.

Huws, Ursula. *The Making of a Cybertariat: Virtual Work in a Real World.* New York: Monthly Review Press, 2003.

Igbaria, Magid, and Margaret Tan, eds. *The Virtual Workplace.* Hershey, PA: Idea Group Publishing, 1998.

Information Technology Association of Canada (ITAC). *Meeting the Skills Needs of Ontario's Technology Sector: An Analysis of the Demand and Supply of IT Professional Skills.* Ottawa: IDC Canada and Aon Consulting, April 2002.

Information Week Research. *2002 National IT Salary Survey.* New York: IWR, 2002.

Itzin, Catherine, and Janet Newman, eds. *Gender, Culture and Organizational Change: Putting Theory into Practice.* London: Routledge, 1995.

Laubacher, Robert, and Thomas Malone."Retreat of the Firm and Rise of Guilds: The Employment Relationship in an Age of Virtual Business." *MIT Initiative on Inventing the Organizations of the Twenty-first Century Working Paper.* Cambridge, MA: MIT, 2002.

Lewis, Theodore. *The Friction-Free Economy: Marketing Strategies for a Wired World.* New York: HarperBusiness, 1997.

Little, Don. *Employment and Remuneration in the Services Industries Since 1984.* Ottawa: Statistics Canada, Services Division, 1999.

Loader, Brian, ed. *Cyberspace Divide: Equality, Agency, and Policy in the Information Society.* New York: Routledge, 1998.

Lopez-Bassols, Vladimir. *ICT Skills and Employment.* Paris: Organization for Economic Cooperation and Development, 2002.

Luxton, Meg. *More Than a Labour of Love: Three Generations of Women's Work In the Home.* Toronto: Women's Educational Press, 1980.

MacIvor, Heather. *Women and Politics in Canada.* Peterborough, ON: Broadview Press, 1996.

MacMurchy, Marjory. *The Canadian Girl at Work: A Book of Vocational Guidance.* Toronto: A.T. Wilgress, 1919.

Maddock, Su. *Challenging Women: Gender, Culture and Organization.* London: Sage, 1999.

Margolis, Jane, and Allan Fisher. *Unlocking the Clubhouse: Women in Computing Education.* Cambridge, MA: MIT Press, 2002.

Massey, Doreen. "Masculinity, Dualisms, and High Technology." In Nancy Duncan, ed., *BodySpace: Destabilizing Geographies of Gender and Sexuality.* 109–126. New York: Routledge, 1996.

Menzies, Heather. *Whose Brave New World? The Information Highway and the New Economy.* Toronto: Between the Lines, 1996.

Miller, Laura. "Women and Children First: Gender and the Settling of the Electronic Frontier." In James Brook and Iain Boal, eds., *Resisting the Virtual Life: The Culture and Politics of Information.* San Francisco: City Lights Books, 1995.

Mitter, Swasti, and Sheila Rowbotham, eds. *Women Encounter Technology: Changing Patterns of Employment in the Third World.* New York: Routledge, 1995.

Morritt, Hope. *Women and Computer Based Technologies: A Feminist Perspective.* Lanham MD: University Press of America, 1997.

Napier, JoAnn, Denise Shortt, and Emma Smith. *Technology with Curves: Women Reshaping the Digital Landscape.* Toronto: HarperCollins, 2000.

Ng, Roxana. "Homeworking: Dream Realized or Freedom Constrained: The Globalized Reality of Immigrant Garment Workers." *Canadian Woman Studies/les cahiers de la femme* 19, no. 3 (Fall 1999): 110–114.

Nicolson, Paula. *Gender, Power, and Organization: A Psychological Perspective.* New York: Routledge, 1996.

Nippert-Eng, Christena. *Home and Work: Negotiating Boundaries Through Everyday Life.* Chicago: University of Chicago Press, 1996.

Noble, David. *Progress without People: New Technology, Unemployment, and the Message of Resistance.* Toronto: Between the Lines Press, 1995.

Organization for Economic Cooperation and Development. *Technology, Productivity and Job Creation.* Paris: OECD, 1996.

Oncu, Ayse Nur. "Informal Economy Participation of Immigrant Women in Canada." MA thesis, University of Alberta, 1992.

O'Reilly, Jacqueline, and Colette Fagan. *Part-Time Prospects: An International Comparison of Work In Europe, North America, and the Pacific Rim.* London: Routledge, 1998.

Orser, Barbara, and Mary Foster. *Home Enterprise: Canadians and Home-Based Work.* Abbotsford, BC: The Home-Based Business Project Committee, 1992.

Pierson, Ruth Roach, and Marjorie Griffin Cohen, eds. *Canadian Women's Issues.* Volume 2. *Bold Visions: Twenty-Five Years of Women's Activism in English Canada.* Toronto: James Lorimer, 1995.

Perelman, Michael. *Class Warfare in the Information Age.* New York: St. Martin's Press, 1998.

Picot, Garnett, and Andrew Heisz. *The Performance of the 1990s Canadian Labour Market.* Ottawa: Statistics Canada, Business and Labour Market Analysis Division, 2000.

Plant, Sadie. *Zeroes and Ones: Digital Women and the New Technoculture.* New York: Doubleday, 1997.

Pollert, Anna. "Dismantling Flexibility." *Capital and Class* 34 (Spring 1988): 42–75.

Pound, Daniel. *Political Economy and Ideology in the Managerial-Technological Society.* Dubuque, IW: Kendall Hunt, 1990.

Prabhu, Sirish. *The Software Development and Computer Services Industry: An Overview of Developments in the 1990s.* Ottawa: Statistics Canada Services Division, 1998.

Probert, Belinda, and Bruce W. Wilson, eds. *Pink Collar Blues: Work, Gender and Technology.* Melbourne: Melbourne University Press, 1993.

Pujol, Michele. *Feminism and Anti-Feminism in Early Economic Thought.* London: Edward Elgar Publishing, 1992.

Redclift, Nanneke, and Thea Sinclair, eds. *Working Women: International Perspectives on Labour and Gender Ideology.* New York: Routledge, 1991.

Romero, Mary, and Elizabeth Higginbotham, eds. *Women and Work: Exploring Race, Ethnicity, and Class.* Thousand Oaks, CA: Sage Publications, 1997.

Sangster, Derwyn. *Assessing and Recognizing Foreign Credentials in Canada: Employers' Views.* Ottawa: Citizenship and Immigration Canada, HRDC, Canadian Chamber of Commerce, and Canadian Labour and Business Centre, 2001.

Schenk, Christopher, and John Anderson, eds. *Re-Shaping Work: Union Responses to Technological Change.* Ontario: Ontario Federation of Labour Technology Adjustment Research Programme, 1995.

Shade, Leslie Regan. *Report on the Use of the Internet in Canadian Women's Organizations.* Ottawa: Status of Women Canada, 1996.

Smith, Ekuwa, and Andrew Jackson. *Does a Rising Tide Lift All Boats? The Labour Market Experiences and Incomes of Recent Immigrants.* Ottawa: Canadian Council on Social Development, 2002.

Spender, Dale. *Nattering on the Net: Women, Power and Cyberspace.* Toronto: Garamond Press, 1995.

Stabile, Carol. *Feminism and the Technological Fix.* New York: St. Martin's Press, 1994.

Stephen, Jennifer. *Access Diminished: A Report on Women's Training and Employment Services in Ontario.* Toronto: Advocates for Community Based Training and Education for Women, June 2000.

Stewart Millar, Melanie. *Cracking the Gender Code: Who Rules the Wired World?* Toronto: Second Story Press, 1998. Now available from Sumach Press, Toronto.

Terry, Jennifer, and Melodie Calvert, eds. *Processed Lives: Gender and Technology in Everyday Life.* New York: Routledge, 1997.

Vickers, Jill, Pauline Rankin, and Christine Appelle. *Politics as if Women Mattered: A Political Analysis of the National Action Committee on the Status of Women.* Toronto: University of Toronto Press, 1993.

Vosko, Leah. *Temporary Work: The Gendered Rise of a Precarious Employment Relationship.* Toronto: University of Toronto Press, 2000.

Wajcman, Judy. *Feminism Confronts Technology.* University Park: Pennsylvania University Press, 1991.

—. *Managing Like a Man: Women and Men in Corporate Management.* University Park: Penn State Press, 1998.

—. *Technofeminism.* Cambridge, UK: Polity Press, 2004.

Wannell, Ted, and Jennifer Ali. *Working Smarter: The Skill Bias of Computer Technologies.* Ottawa: Statistics Canada and HRDC, May 2002.

Waring, Marilyn. *Counting for Nothing: What Men Value and What Women Are Worth.* Toronto: University of Toronto Press, 1999.

Webster, Juliet. *Office Automation: The Labour Process and Women's Work in Britain.* New York: Harvester Wheatsheaf, 1990.

Willcocks, Leslie, and Stephanie Lester, eds. *Beyond the IT Productivity Paradox.* Toronto: John Wiley and Sons, 1999.

Wright, Barbara Drygulski, ed. *Women, Work, and Technology: Transformations.* Ann Arbor: University of Michigan Press, 1987.

Zuboff, Shoshana. *In the Age of the Smart Machine: The Future of Work and Power.* New York: Basic Books, 1988.

INDEX

ability 21, 32
age 17, 21, 22, 53–54, 136
Alliance@IBM 162
Andrea 110
Angie 75–76
Ann 93–94
Anya 102

Barbados 113, 155-156, 160
benefits 14, 38, 113, 116, 142, 145, 162
Brenda 31–32, 36, 40, 56–57, 92

call centres 27; location of 26, 120; work in 17, 106–108, 120–121
career path 18, 35, 51, 54, 90, 91, 96, 174
caregiving 50, 52, 116, 121, 148; to children 35-37, 39, 40, 52, 60, 77, 117, 140, 141, 144, 151, 175; to ill and elderly 50, 148
Carmen 55–56, 57, 101–102
certification *see* credentialization
Charity 45, 89
Charmaine 74
childlessness 48
Cisco 14, 93
citizenship, technological 16
class 17, 19, 21, 28, 32, 35, 66
clerical occupations 17, 80–82, 85, 87, 121, 126, 133, 142
Cockburn, Cynthia 166
credentialization 27, 65, 70, 73, 77, 84–85, 90, 95–96, 103–104, 114-115, 120, 123–125, 133; and private IT instruction 114, 123–124
cyberfeminism 166
Cynthia 155–156

data entry 46, 112, 141–142
Debbie 94
Diane 38, 41, 45, 120, 155, 156–157
discrimination 17, 72, 96, 167
domestic work, paid 35, 47
domestic work, unpaid *see* unpaid labour
Donna 47, 178–179
double day 35
Dryburgh, Heather 53
Dyer-Witheford, Nick 108–109, 112

earnings: annual general IT salaries 14, 28, 38, 68, 80, 110, 120, 122, 123, 145; inequity in 17, 19, 20, 27, 34, 51, 53, 59, 60, 79-80, 175, 180-183, 206 n.37
education, computer science or IT programs 15, 54, 89, 91, 115, 126; general attainment 70, 99, 123; statistics on 15, 91, 122, 123, 181; women's participation in 15, 18, 31, 69, 77–79, 90–91, 181, 183
electronic cottage 40, 48–49
employment: contract 13, 15, 18, 37, 45, 70, 106, 131, 141–142; freelance 15, 37, 45, 131, 140–142, 151;

full-time 37, 141–142; non-standard 37, 132–134; part-time 15, 18, 37, 45, 70, 120, 131, 141; standard 37, 132–133; temporary 37, 70, 131
Erin 55, 89–90, 91, 97–98, 98–99, 186

Faith 152, 154–155
Farideh 177–178
Fast Company magazine 12, 109
father's influence 88–90
femininity, ideologies of 41–44, 48–49, 57, 83, 95, 116–118, 121, 160
feminism 174–179
feminist research 26, 35
feminization of employment 56, 80, 113, 120, 134, 161
flexibility: for employees 37–39, 47, 119, 131, 134–135, 137, 144, 185-186; for employers 37, 134, 135, 137, 144
Fordism 219 n.4
Franklin, Ursula 61, 85, 103, 104, 174
Freeman, Carla 113–114, 160
friction-free economy 11
Friedan, Betty 41–43, 45, 47, 49, 204 n.12
futurism 11,12, 49, 61

Gillian 175–176
globalization 13, 113, 133, 157, 160
government 14, 16
harassment, workplace or sexual 17
Haraway, Donna 159
Helen 41, 75
Hewlett-Packard 14, 109, 174
homework 14, 36–41, 48–49, 130–131, 140–145; history of 48, 145; to manage domestic responsibilities 36, 39, 44, 47, 144; valuation of, gendered 17–18, 39, 40, 41
household technologies 42, 61
housework *see* unpaid labour
human capital 67, 69–73, 76–78, 79, 81
Human Resources Development Canada 16, 67, 108, 117
hybrid jobs 15, 56, 102

IBM 14, 109, 162
identities: complexity of 21–24; markers of 18, 20
immigrants 20, 35, 46, 53, 56, 59, 63–66, 72–73, 87, 103, 135, 156, 173
India 112–113, 158–160, 172, 187
industrial segregation 35, 52, 58, 59, 79, 82, 160
innovation 16, 17, 49, 67, 114
IT industry: changes in 11–14, 17–18, 68, 108–112; culture of 108–112; defining 15, 17, 27, 28; gender composition of 14, 15, 17–18, 51, 57; history of 13, 14, 28

Jamie 102
Jane 151
Janice 47, 130

Karen 45, 132
Katherine 39
knowledge community 21, 58, 103–104, 108
knowledge economy, knowledge workers *see* knowledge work
knowledge work 16, 17, 26, 65, 67, 70, 83–84, 103, 112, 127

Laurie 55, 86–88, 124–125, 151, 152
leisure time 11, 147
Lilith 55, 76, 84–85, 88, 149–150, 151, 185, 187
Lina 63–64
Lisa 74–75

Maggie 145
Maria 53–54, 153
Maritimes 17, 113, 142, 157
Mary 151
masculinity: ideologies and communities of 20, 57–58, 69, 92–93, 96, 99, 109
materialist perspective 23, 62
Melanie 47, 183
Menzies, Heather 50, 141, 166
Mike 106–108
monitoring of work, of employees *see* technological surveillance
Montreal 107
motherhood 21, 39, 76, 151–152

Nadine 55
Naomi 152–153
neo-familialism 148–149
Neetha 151, 172–174, 179
Nortel 14, 93, 112, 169–170, 201 n.10

occupational classification 28
occupational segregation 18, 20, 34, 51–52, 55–56, 58–59, 79, 82, 160
Ontario 25, 67, 70, 108, 119, 121
outsourcing 14, 72, 113, 130, 136, 143, 157, 160, 171

Patricia 138–140
pay equity *see* earnings, inequity in
Penny 89
Plant, Sadie 98
post-Fordist 132, 219 n.4
Preeti 92, 97
productivity 52, 76, 147–148
progress, technological 70, 112, 134, 170

race/ethnicity 17, 19–22, 28, 32, 66, 106
Rebecca 135–138
regional disparity 26, 54, 79, 113, 118, 142–143, 157
Rogers 129, 218 n.1

salaries *see* earnings
Sandra 143–144
Sarah 60, 94–95
self-employment 13, 18, 37–39, 40, 41, 47, 103, 120, 141, 145, 151
seniority 51, 56
service industries, general 52, 70, 113, 116–119
service industries, IT-related 116, 118–119
sexuality 17, 21, 32, 156-157
shadow work 50
Shiva, Vandana 166
Sitel 158
skills 65, 66; commodification of 67, 71, 103, 104, 115, 124; compensation for 60, 66, 68, 70, 79–81, 96, 103; deskilling 27, 81–83, 115; immigrants and 63–66, 72-73; perception of 16, 17, 27, 28, 56, 58, 69, 70, 73-76, 86, 96, 99, 103, 125; perception of, gendered 17, 28, 74, 77, 81, 100–101, 116–118, 152, 154; and personal empowerment 66–67, 70, 94, 103, 167–168, 175, 179; and service industries 116–118; shortages and deficits 65, 67, 68, 85, 96, 114–116, 124; upskilling 81–83; technical and non-technical (or hard and soft skills) 17, 27, 57–58, 83, 99–102
social location 19, 32–34, 66, 72
socialization, gendered 73–74; and parental support 88, 89
space, domestic 49

Sprint 129
Statistics Canada 19, 28
Stephanie 78
stress 14
structural relations 19, 21–24, 28, 33–35, 59, 61–62, 72, 120, 146, 148, 171
subcontracting 27, 49, 129, 142, 145
Susan 53

Tari 105–106, 110–111, 120, 123–124
Tatiana 75, 187
techiladoras 160, 223 n.43
technological change in the workplace 61, 81, 82, 133, 146
technological surveillance 42, 116–117, 165
telework 139–144; and telecommuting 141–143
Toffler, Alvin 49
Toronto 25, 26, 54, 107, 130, 139, 164, 172
training: availability of and access to 77–78, 85–86, 93-96, 115, 173; benefits of 80, 85; for call-centre work 107

unions 14, 20, 34, 58, 104, 130, 145, 162
United States 157–158, 161
unpaid labour 17, 35, 39, 40, 42, 44, 48, 71, 141, 175; gender divisions of 34–35, 50, 52, 148; and restructuring 50, 148–149

video games 75–76
visible minorities 53, 60, 64

wages *see* earnings
WashTech 162
Wild West or frontier metaphor 109, 114
Wired magazine 12, 109, 159, 187
Women in Global Science and Technology (WIGSAT) 179–180
women of colour *see* visible minorities
women's technology organizations 74–75, 167–168, 171, 174–177
Woolf, Virginia 43, 47
work: boundaries of 14, 33–34, 36–37, 39, 44–45, 139–141, 146–147, 149; choice of 17,18, 22, 24, 28, 35, 37, 71, 149; fragmentation of 12, 42, 61, 85, 114, 135; location of 13, 26, 34, 43–44, 129–130, 139–141, 143, 146, 160; space of, defining 36, 39–45, 49, 130, 140; speed of 126, 137–138; quality of 11–14, 19, 25, 27, 32, 144, 171, 175, 181, 184, 188; valuation of 40–42, 52, 56–58
work arrangements and hours of 14, 37–39, 45, 130–132, 136, 144, 147, 149–152, 185
work at home *see* homework
workplace, climate and culture of 20, 56, 91, 95–96, 137, 153–155
WorldCom 14, 112, 170

Y2K 13, 106–107, 216 n.1

Other titles from The Women's Issues Publishing Program

Troubling Women's Studies: Pasts, Presents and Possibilities
Ann Braithwaite, Susan Heald, Susanne Luhmann and Sharon Rosenberg

Inside Corporate U: Women in the Academy Speak Out
Edited by Marilee Reimer

Out of the Ivory Tower: Feminist Research for Social Change
Edited by Andrea Martinez and Meryn Stuart

Strong Women Stories: Native Vision and Community Survival
Edited by Kim Anderson and Bonita Lawrence

Back to the Drawing Board: African-Canadian Feminisms
Edited by Njoki Nathane Wane, Katerina Deliovsky and Erica Lawson

Cashing In On Pay Equity? Supermarket Restructuring and Gender Equality
Jan Kainer

Double Jeopardy: Motherwork and the Law
Lorna A. Turnbull

Turbo Chicks: Talking Young Feminisms
Edited by Allyson Mitchell, Lisa Bryn Rundle and Lara Karaian
Winner of the 2002 Independent Publishers Award

Women in the Office: Transitions in a Global Economy
Ann Eyerman

Women's Bodies/Women's Lives: Women, Health and Well-Being
Edited by Baukje Meidema, Janet Stoppard and Vivienne Anderson

A Recognition of Being: Reconstructing Native Womanhood
Kim Anderson

Women's Changing Landscapes: Life Stories from Three Generations
Edited by Greta Hofmann Nemiroff

Women Working the NAFTA Food Chain: Women, Food and Globalization
Edited by Deborah Barndt, Winner of the 2000 Independent Publishers Award

Cracking the Gender Code: Who Rules the Wired World?
Melanie Stewart Millar, Winner of the 1999 Independent Publishers Award

Redefining Motherhood: Changing Identities and Patterns
Edited by Sharon Abbey and Andrea O'Reilly

Fault Lines: Incest, Sexuality and Catholic Family Culture
Tish Langlois